SpringerBriefs in Electrical and Computer Engineering

More information about this series at http://www.springer.com/series/10059

Deyu Zhang · Zhigang Chen
Haibo Zhou · Xuemin (Sherman) Shen

Resource Management for Energy and Spectrum Harvesting Sensor Networks

Springer

Deyu Zhang
School of Software
Central South University
Changsha, Hunan
China

Zhigang Chen
School of Software
Central South University
Changsha, Hunan
China

Haibo Zhou
Department of Electrical and Computer
Engineering
University of Waterloo
Waterloo, ON
Canada

Xuemin (Sherman) Shen
Department of Electrical and Computer
Engineering
University of Waterloo
Waterloo, ON
Canada

ISSN 2191-8112   ISSN 2191-8120 (electronic)
SpringerBriefs in Electrical and Computer Engineering
ISBN 978-3-319-53770-2   ISBN 978-3-319-53771-9 (eBook)
DOI 10.1007/978-3-319-53771-9

Library of Congress Control Number: 2017931571

Printed on acid-free paper

This Springer imprint is published by Springer Nature
The registered company is Springer International Publishing AG
The registered company address is: Gewerbestrasse 11, 6330 Cham, Switzerland

Preface

To alleviate the energy and spectrum constraints in wireless sensor networks (WSNs), WSNs necessitate energy and spectrum harvesting (ESH) capabilities to scavenge energy from renewable energy sources, and opportunistically access the underutilized licensed spectrum; hence, give rise to the energy and spectrum harvesting sensor networks (ESHSNs). In spite of the energy and spectrum efficiency brought by ESHSNs, their resource management faces new challenges. First, energy harvesting (EH) process is dynamic, which makes balancing energy consumption and energy replenishment challenging. Depleting the sensors battery at a rate slower or faster than the energy replenishment rate leads to either energy underutilization or sensor failure, respectively. Second, the spectrum utilization by sensors in ESHSNs has to adapt to the dynamic activity of primary users (PUs) over licensed spectrum.

In this monograph, we investigate the resource management and allocation, to facilitate energy- and spectrum-efficient sensed data collection in ESHSNs. In Chapter 1, we discuss the motivation to integrate ESH capabilities in WSNs, as well as the network architecture, typical application scenarios, and challenges of ESHSNs. Chapter 2 surveys the related state-of-the-art research literature. In Chapter 3, an EH-powered licensed spectrum sensing scheme is proposed to schedule the spectrum sensors which are dedicatedly deployed for spectrum sensing to periodically estimate the licensed spectrum availability. Accordingly, an access time and power management of battery-powered data sensors is presented as well, which has been verified an effective solution to minimize the energy consumption of data transmission over the available licensed spectrum. In Chapter 4, we propose an online algorithm which jointly manages the available licensed spectrum and harvested energy to optimize the network utility which captures the data collection efficiency of ESHSNs. The proposed algorithm dynamically schedules sensors' data sensing and spectrum access by considering the stochastic nature of EH process, PU activities, and channel conditions. Finally, Chapter 5 concludes the monograph by outlining some open issues, pointing out new research directions for resource management in ESHSNs.

The authors would like to thank Ning Zhang of the BroadBand Communications Research (BBCR) group at University of Waterloo, Prof. Mohamad Khattar Awad of Kuwait University, and Prof. Ju Ren of Central South University for their contribution to the presented research works, and Prof. Shibo He of Zhejiang University for his valuable suggestions on the monograph draft. We also would like to thank all the members of BBCR group for their valuable comments and suggestions. Special thanks are due to the staff at Springer Science+Business Media: Susan Lagerstrom-Fife and Jennifer Malat, for their help throughout the publication preparation process.

Changsha, China   Deyu Zhang
Changsha, China   Zhigang Chen
Waterloo, ON, Canada   Haibo Zhou
Waterloo, ON, Canada   Xuemin (Sherman) Shen

Contents

Acronyms

AoI	Area of Interest
CR	Cognitive Radio
CSMA/CA	Carrier Sense Multiple Access with Collision Avoidance
EH	Energy Harvesting
EMI	ElectroMagnetic Interference
ESHSN	Energy and Spectrum Harvesting Sensor Network
ESI	Energy State Information
HSHSN	Heterogeneous Spectrum Harvesting Sensor Network
ISM	Industrial, Scientific, and Medical
MAC	Media Access Control
MDP	Markov Decision Process
PU	Primary User
QoS	Quality of Service
RF	Radio Frequency
SH	Spectrum Harvesting
SHSN	Spectrum Harvesting Sensor Network
SNR	Signal-to-Noise Ratio
SoC	State of Charge
SP	Spectrum Provider
SU	Secondary User
TDMA	Time Division Multiple Access
TPS	Third-Party System
WSN	Wireless Sensor Network

Chapter 1
Introduction

Although wireless sensor networks (WSNs) have been widely deployed to perform monitoring and surveillance tasks, their performance is highly deteriorated by energy and spectrum scarcities. By empowering sensors with energy and spectrum harvesting (ESH) capabilities, the emerging ESH sensor networks (ESHSNs) can access the idle licensed spectrum using harvested energy, and thus fundamentally address the resource constraint issues. In this chapter, we present the network architecture and several key applications of ESHSNs. Then we discuss the challenges of resource allocation in ESHSNs and the aim of this monograph.

1.1 Resource Constraints in Wireless Sensor Networks

With the development of sensing technologies, low-power embedded systems, and wireless communication, WSN has become a promising solution to collect information and assist users (e.g., machines or humans) for interaction with real-world objects. Nowadays, sensors have permeated more and more aspects of personal, including vehicles, washing machines, air conditioners, etc. According to a report from Frost & Sullivan, the global market of WSNs is forecast to increase from 1.4 billion to 3.26 billion during 2014–2024 [1].

Generally, a WSN consists of miniaturized and low-end sensors powered by batteries of limited capacity. To guarantee the long-term operation, an operator needs to manually replace the depleted battery, which results in considerable maintenance cost. Numerous energy-efficient schemes are proposed to reduce sensors' energy consumption, such as compressive sensing, cooperative multiple-input-multiple-output (MIMO) and energy-efficient MAC/network layer protocols [2]. However, the difficulty remains for long-term unintended operation of WSNs due to the limited battery capacity, which is referred to as the energy scarcity problem.

© The Author(s) 2017   1
D. Zhang et al., *Resource Management for Energy and Spectrum
Harvesting Sensor Networks*, SpringerBriefs in Electrical
and Computer Engineering, DOI 10.1007/978-3-319-53771-9_1

In addition, sensors are networked through the free-of-charge unlicensed spectrum, i.e., the industrial, scientific, and medical (ISM) spectrum. Although the utilization of unlicensed ISM spectrum facilitates the explosive deployments of WSNs, their data transmissions have been heavily interfered by other unlicensed wireless devices, e.g., Wi-Fi networks and Bluetooth [3]. According to a report of Consumer Electronic Association [4], the number of unlicensed wireless devices has increased by threefolds in the past 8 years; and the growth continues at an annual rate of 30%. Due to the proliferation of unlicensed devices, more and more devices tend to coexist with WSNs, which bring significant interferences to the latter. This problem is referred to as the spectrum scarcity problem.

1.2   Enabling Techniques for Energy and Spectrum Harvesting

The resource scarcities significantly degrade the lifetime and data collection performance of WSNs. To mitigate these issues, WSNs urgently demand new supply of energy and spectrum to improve the network performance. In this section, we introduce the enabling techniques for energy and spectrum harvesting (ESH) which enable sensors to exploit the energy harvested from ambient energy resources and underutilized licensed spectrum.

1.2.1   Energy Harvesting

Energy harvesting (EH) enables sensors to convert the energy in ambient renewable energy sources to electric power, such as solar and wind in outdoor scenarios, and vibration and heat in industrial environments. Continuous advances in EH technology improve the miniaturization of EH equipments and make them more applicable in small-size sensors. Relying on the widely existing ambient energy sources in the area of interest (AoI), the energy supply brought by EH can be deemed as infinity; thus, EH fundamentally addresses the energy scarcity problem. With a well-designed energy management scheme, the EH-powered WSNs (EHWSNs) can achieve perpetual operation time and provision high quality of service (QoS).

1.2.2   Spectrum Harvesting

Although the unlicensed spectrum has been over-crowded, it is observed by Federal Communications Commission that a large portion of the spectrum licensed to primary users (PUs) is merely sporadically used [5]. To mitigate this resource unbalancing, the cognitive radio (CR) emerges to enable the unlicensed devices to sense the surrounding electromagnetic environments and adjust the operating parameters to access the spectrum licensed to other users [6].

By equipping CR, sensors can explore and exploit the temporally available licensed spectrum for data transmissions and thus become capable to harvest energy to alleviate the spectrum scarcity problem [7]. Furthermore, since sensors can select the spectrum with less fading and interference, SH capability potentially improves the energy efficiency of WSNs. Notably, sensors access licensed spectrum as secondary subscribers. Their data transmission cannot hinder the activity of the spectrum owners, i.e., the PUs. Therefore, different from the unlicensed spectrum allocation in traditional WSNs, sensors may frequently vacate and change the operating spectrum to avoid interferences to PUs.

1.3 Energy and Spectrum Harvesting Sensor Networks

The integration of sensors with ESH capabilities gives birth to energy and spectrum harvesting sensor networks (ESHSNs) which operate over idle licensed spectrum using harvested energy. This section discusses the network architecture, the key applications, and the research challenges of ESHSNs.

1.3.1 Network Architecture

Figure 1.1 illustrates the network architecture of an ESHSN that consists of a sink and numerous sensors with ESH capabilities (i.e., ESH sensors). The ESHSN coexists with a primary network which has the privilege to access the licensed spectrum. The sensors use harvested energy to sense data from the AoI, and then transmit the sensed data to the sink through the idle licensed spectrum, as shown by green links in Fig. 1.1. To avoid the interference to PUs, the spectrum detection is mandatory, which can be realized by either a third-party system (TPS) or the sensors themselves. Other than sensed data, the sensors may exchange auxiliary information with each other and the sink for spectrum handoff, sensing rate control, and routing, etc.

Figure 1.2 describes a typical ESH sensor, which consists of five modules: the EH module, the rechargeable battery, the data sensing/processing module, the data buffer, and the CR transceiver, respectively. The EH module harvests energy from ambient energy sources and store the energy in the rechargeable battery. The stored energy can be used by the data sensing/processing module which is responsible to collect information from the AoI, and the CR transceiver which detects the licensed spectrum, and receives and transmits data over the spectrum. The received and sensed data are stored in the data buffer.

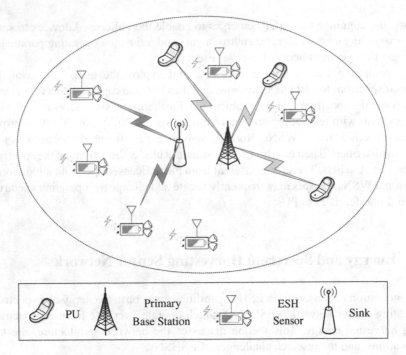

Fig. 1.1 Network architecture of an ESHSN

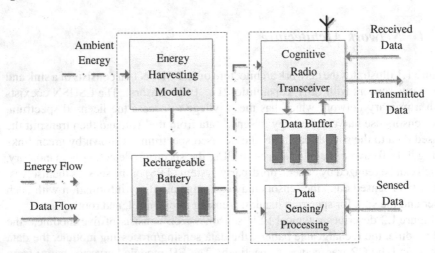

Fig. 1.2 Diagram of an ESH sensor

The ambient energy sources to harvest from determine the type of the EH module. For example, the energy in sunlight can be harvested by solar panel, and the energy in changes of ambient conditions, such as pressure, temperature, and acceleration, can be transformed into electric energy by piezoelectric transducers. To store harvested energy and prevent the impacts of batteries' memory effect,[1] the super capacity or lithium-ion battery are considered as competitive candidates for rechargeable battery in the ESH sensors.

Considering the limitation on the manufacture cost of sensors, the low-cost CR transceivers are desired, which can cover the licensed spectrum of interest, i.e., the spectrum that is accessible for ESH sensors. Since the price of the CR transceivers tends to increase with the width of the tunable frequency range, it is important to choose the CR transceiver according to the accessed licensed spectrum. For example, given that the ESH sensors can access the 700 MHz or 800 MHz spectrum that is licensed to cellular users, the RTL-SDR with spectrum range 24–1766 MHz and price around $20 [8] is affordable for the ESH sensors.

1.3.2 Applications of ESHSNs

With ESH capabilities, ESHSNs can perform monitoring and surveillance tasks sustainably while overlapping with other unlicensed networks, such as Wi-Fi hotspots and Bluetooth. Therefore, ESH capabilities significantly improve the applicability of WSNs in various real-life scenarios:

1.3.2.1 Smart City Applications

The emerging smart city requires advanced sensors that are integrated in the real-time monitoring systems, including transportation system and power grid, to intelligently manage city's assets and tackle inefficiency. However, the wireless channel condition on the ISM spectrum is quite harsh and perverse due to the complex propagation environment brought by densely situated buildings. On the other hand, large portions of the licensed spectrum are underutilized even in the populated urban area, especially the TV white space [9, 10]. SH capability enables ESH sensors to access the temporally available spectrum, and hence decrease data transmission delay in real-time monitoring. Furthermore, ambient energy sources widely exist in the urban area, such as solar, wind, and radio frequency (RF), from which sensors can achieve sustainable operation without human intervention.

---

[1] An effect that the rechargeable batteries gradually lose their maximum energy capacity, if they are repeatedly recharged after being partially discharged.

1.3.2.2 Indoor Surveillance and Localization Applications

The indoor scenarios, such as houses or factories, require a large number of sensors for surveillance or localization. Nowadays, plenty of indoor scenarios have been covered by Wi-Fi hotspots to satisfy the rapid increasing demand for convenient Internet access. Since the transmission power of Wi-Fi hotspots is much higher than the power of sensor networks, the coexistence of hotspots and sensor networks significantly degrades the network performance of the latter. The interference makes the data transmission rate of the sensor networks to decrease by nearly 20%, as reported in [11].

ESHSN can avoid the inter-network interference by dynamically exploiting the underutilized licensed spectrum, and thus becomes more flexible to complex spectrum environment in indoor scenarios. Moreover, with artificial illumination and thermal as the renewable energy sources, ESHSN alleviates the stringent constraint of energy supply. Therefore, ESHSN is able to extend the network lifetime and flexibly accomplish surveillance tasks in indoor scenarios.

1.3.2.3 Electronic Health Applications

Electronic health (E-health) system consists of body sensors implanted in/on the patient's body to monitor the health information including pressure, blood oxygen, etc. The information is collected by the central controller for prevention or early detection of diseases. Due to the interference and congestion brought by other unlicensed networks in the context of E-health system, the ISM spectrum becomes no longer sufficient for life-critical health monitoring. In addition, the operation of body sensors also needs to take into account the electromagnetic interference (EMI) to the bio-medical devices, such as electrocardiograph monitor and electromyography. The opportunistic spectrum access enables ESHSNs to leverage larger range of spectrum, and avoid the EMI to ambient bio-medical devices.

Furthermore, the sustainable operation is another critical concern of body sensors, where the battery replacement may require surgical procedures. ESHSNs can harvest energy from the human body, such as kinetic and mechanic movements to continuously monitor the critical health information.

1.3.3 Challenges for ESHSNs

Unlike the stable supply of battery-saved energy and fixed spectrum allocation in traditional WSNs, the availability of both harvested energy and spectrum is subject to conditions of the ambient energy sources and PU activities, respectively. The efficient utilization of these uncontrollable resources poses much more challenges to the scheduling of ESHSNs. In the following, we discuss the challenges from the perspective of the coupling of energy and spectrum allocation, and the stochastic nature of EH process and PU activities.

1.3.3.1 Coupling of Energy and Spectrum Allocation

To guarantee the energetic sustainability and PU protection, ESH sensors' data sensing and transmission need to adapt to the supply of harvested energy and spectrum. The allocation of the two resources jointly affects the network performance. For example, the sensors may have limited chances for data transmission to avoid collisions in case of frequent PU activities on the licensed spectrum. Even with sufficient supply of harvested energy, the sensors should not collect too much data to avoid data buffer overflow. Furthermore, the tuning of energy consumption in data sensing and transmission can be modeled by a continuous variable, while the allocation of spectrum can be modeled by integer variables. Therefore, the joint allocation of energy and spectrum falls into the class of mixed integer programming problem, which is in general difficult to solve.

1.3.3.2 Handling of Multiple Stochastic Processes

In addition to the stochastic nature in EH process and PU activities, the quality of licensed channels also exhibits stochasticity due to time-varying and stochastic conditions, such as multipath propagation, shadowing, etc. Therefore, another challenge for resource allocation in ESHSNs is how to deal with the multiple stochastic processes in energy charging, PU activities, and channel conditions. The associated resource allocation problem belongs to the class of stochastic optimization.

Markov decision process (MDP) has been widely used in existing works to handle the stochasticity in EH process, PU activities and channel fading, with the objective to improve the network throughput [12, 13]. However, in typical sensor network applications, sensors are densely deployed to collect data and transmit the data to one or several sinks. The number of sensors is usually much larger than that of the sinks, which constitutes the many-to-one traffic pattern. This pattern makes the MDP-based solution not practical for the ESHSN, because the complexity of MDP exponentially increases with the number of sensors [14, 15].

1.4 Aim of the Monograph

By empowering sensors with ESH capabilities, ESHSNs can operate on the underutilized licensed spectrum using harvested energy, liberating sensor networks from the resource constraint problems. However, the stochastic availability of harvested energy and spectrum make the resource allocation issues in this new sensor networking paradigm more challenging.

In this monograph, we attempt to investigate the utilization of harvested energy and idle licensed spectrum for ESHSNs, with the objective to improve the network performance while guaranteeing the PU protection and sensors' sustainability. Specifically, we make an effort to address the following research issues: (a) how to identify the underutilized licensed spectrum by exploiting harvested energy and (b) how to deal with the stochasticity in EH process and PU activities. To answer these questions, we investigate the scheduling of EH-powered sensors to maximize the detected spectrum available time, and network utility optimization jointly considering the stochastic processes and spectrum detection error. Based on the investigation on these issues, we can elaborate the insights and implications for design of ESHSNs.

References

1. http://www.smartgridnews.com/story/major-growth-forecast-wireless-sensor-market/2015-01-28
2. N.A. Pantazis, S.A. Nikolidakis, D.D. Vergados, Energy-efficient routing protocols in wireless sensor networks: a survey. IEEE Commun. Surv. Tutorials **15**(2), 551–591 (2013)
3. A. Ahmad, S. Ahmad, M. Rehmani, N. Ul Hassan, A survey on radio resource allocation in cognitive radio sensor networks. IEEE Commun. Surv. Tutorials (to be published)
4. Consumer Electronics Association, Unlicensed spectrum and the U.S. economy quantifying the market size and diversity of unlicensed devices (2014)
5. I.F. Akyildiz, W.-Y. Lee, M.C. Vuran, S. Mohanty, Next generation/dynamic spectrum access/cognitive radio wireless networks: a survey. Comput. Netw. **50**(13), 2127–2159 (2006)
6. I.F. Akyildiz, B.F. Lo, R. Balakrishnan, Cooperative spectrum sensing in cognitive radio networks: a survey. Phys. Commun. **4**(1), 40–62 (2011)
7. O. Akan, O. Karli, O. Ergul, Cognitive radio sensor networks. IEEE Netw. **23**(4), 34–40 (2009)
8. RTL Software Defined Radio, http://www.rtl-sdr.com/roundup-software-defined-radios/
9. A.B. Flores, R.E. Guerra, E.W. Knightly, P. Ecclesine, S. Pandey, IEEE 802.11af: a standard for TV white space spectrum sharing. IEEE Commun. Mag. **51**(10), 92–100 (2013)
10. H. Zhou, N. Cheng, N. Lu, L. Gui, D. Zhang, Q. Yu, F. Bai, X.S. Shen, Whitefi infostation: engineering vehicular media streaming with geolocation database. IEEE J. Sel. Areas Commun. **34**(8), 2260–2274 (2016)
11. Avoiding RF Interference Between WiFi and Zigbee, Technical Report, http://www.mobiusconsulting.com/papers/ZigBeeandWiFiInterference.pdf
12. S. Park, D. Hong, Optimal spectrum access for energy harvesting cognitive radio networks. IEEE Trans. Wirel. Commun. **12**(12), 6166–6179 (2013)
13. X. Lu, P. Wang, D. Niyato, E. Hossain, Dynamic spectrum access in cognitive radio networks with RF energy harvesting. IEEE Wirel. Commun. **21**(3), 102–110 (2014)
14. Z. Wang, V. Aggarwal, X. Wang, Power allocation for energy harvesting transmitter with causal information. IEEE Trans. Commun. **62**(11), 4080–4093 (2014)
15. D. Zhang, Z. Chen, M.K. Awad, N. Zhang, H. Zhou, X.S. Shen, Utility-optimal resource management and allocation algorithm for energy harvesting cognitive radio sensor networks. IEEE J. Sel. Areas Commun. **34**(12), 3552–3565 (2016). doi:10.1109/JSAC.2016.2611960

Chapter 2
Energy and Spectrum Harvesting in Sensor Networks

This chapter provides a comprehensive survey of existing literature of energy and spectrum allocation regarding ESHSNs, to understand the issues related in a better way. We first survey the EH process modeling and energy allocation which provide insights for efficient utilization of harvested energy. Then the spectrum sensing and resource allocation in SHSNs are discussed, which are of great importance to improve spectrum efficiency while guarantee PU protection. At last, we discuss the joint energy and spectrum management in ESHSNs.

2.1 Energy Harvesting

The ambient energy sources widely exist in the area of interest (AoI) of sensor networks, such as solar and wind in outdoor scenarios, vibration and heat in industrial scenarios, etc. [1]. By equipping sensors with EH module, such as solar panel and piezoelectric transducers, they can scavenge energy from the energy sources and continuously operate without battery replacement. In this section, we survey the state-of-the-art research regarding ESHSNs from the perspective of EH process modeling and energy allocation, respectively.

2.1.1 EH Process Modeling

The efficient allocation of energy in ESHSNs relies largely on the accurate model of EH process. Based on the availability of non-casual or causal knowledge of EH process at sensors, the models in the literature can be categorized into two classes, i.e., the deterministic model and stochastic models [1].

© The Author(s) 2017
D. Zhang et al., *Resource Management for Energy and Spectrum Harvesting Sensor Networks*, SpringerBriefs in Electrical and Computer Engineering, DOI 10.1007/978-3-319-53771-9_2

2.1.1.1   Deterministic Models

In deterministic models, sensors have the full knowledge of energy arrival instant and amount in advance, i.e., non-casual knowledge. By having the non-casual knowledge, deterministic models facilitate the design of the optimal energy allocation strategy, and assist the designers to set benchmark of performance limits of EH-powered WSNs.

The success of deterministic models heavily depends on the accurate energy profile prediction. However, the EH process can be impacted by numerous environmental factors, which makes the difficulty remains for fully understanding the behavior of ambient energy sources. The prediction error may lead to nodes failure or energy waste. When the sensors predict a sufficient energy supply in the near future, they tend to consume a large amount of energy on data sensing and transmission to improve the network performance. The prediction error leads to sensors' failure due to battery depletion. On the other hand, sensors become conservative on energy consumption if the predicted energy supply is limited. In this case, the prediction error may lead to battery overflow and wastage of energy. In [2], Wang et al. model the imperfect EH prediction by an independent random variable. In [3], Lee et al. address the prediction errors by a token-bucket-based algorithm. Summarily, the deterministic models are suitable for the applications with accurately predictable energy sources, such as solar power in large time scales.

2.1.1.2   Stochastic Models

Recently, the stochastic models have attracted abundant attention for EH process modeling. Since it does not require the non-causal knowledge of EH process, the stochastic model is adequate for the applications that cannot accurately predict the energy state information (ESI). In [4–8], the authors assume that the energy arrives according to an independent and identical distribution (i.i.d.) with fixed EH rate. Although the i.i.d. EH models capture the intermittent nature of EH process, they cannot characterize the temporal correlation of ambient energy sources, such as solar or human motion.

Numerous works exploit Markov process to model the temporal correlation of EH process [9–13]. Zordan et al. consider a two-state ("GOOD" or "BAD") Markov chain to represent the time-correlated evolution of the energy source state in [9]. If the energy source state is bad, no energy arrives. Otherwise, the energy quantum arrives according to some mass distribution functions. In [10], Niyato et al. use a generalized Markovian model with multiple states, to describe the impact of clouds on the solar intensity. The Markov chain model is suitable for the illustration of some energy sources. For example, the weather states of solar power harvesting may change between cloudy and clear, and harvesting from vehicular engine vibration can be described by two states which represents the vehicle is either in rest or motion. Considering that the sensors may not be able to directly observe the ambient energy source state, [11–13] use hidden Markov chain to characterize the EH process, in which sensors only have the knowledge of the amount of arrived energy, rather than the states of the ambient energy sources.

In addition to the EH process models, an appropriate choice of the underlying parameters, such as the transition probability between states and the mass distribution function, is of significance yet challenging for efficient energy allocation. In [14], Ku et al. adopt a real data-driven hidden Markov chain to capture the dynamics in the empirical solar power data. The model quantifies EH conditions to several representative states, and then sets the transition probabilities by training with historical solar power data.

2.1.2   Energy Allocation

Based on the knowledge of EH processes, sensors can allocate energy for data sensing and transmission to accomplish the operation tasks, such as event detection, source estimation or data collection, etc. Unlike battery-powered sensor networks that have stable energy supply, the energy allocation in ESHSNs needs to balance the energy consumption and energy harvesting. Depleting a sensor's battery at a rate lower or faster than the harvesting rate may lead to either energy overflow or energy outage, respectively. The energy allocation can be casted into three categories, i.e., the static, offline, and online cases, depending on the knowledge of EH process at the sensors. In static case, the sensors only have the EH rate at the current time slot without any non-casual or statistics knowledge. In offline and online cases, sensors have the non-casual or casual knowledge of EH process, respectively. In the following, we discuss the energy allocation in the static case, offline case and online case, respectively.

2.1.2.1   Static Energy Allocation

For static energy allocation, the sensors schedule energy allocation according to the current state of ambient energy sources, and reschedule when the state varies. Since only the current state of energy sources is required, static energy allocation significantly simplifies the prediction and modeling of EH process, making it suitable for applications with slowly variate sources. However, the static energy allocation may not be able to promise the long-term network performance due to the lack of non-casual and statistics knowledge of EH process.

In [15], the authors investigate the energy allocation in an EH-powered body sensor network for medical event detection. To improve the quality of service (QoS), the authors propose a QoS control scheme which schedules event detection and transmission subject to the energy neutrality constraint,[1] and drops the data with limited clinical validity. By exploiting the sensors around the source and the sink as relays,

---

[1]The cumulative energy consumed cannot surpass the cumulative energy harvested by that time [16].

Zhang et al. propose a relay selection and power control scheme for EH-powered sensor networks to balance the residual energy in sensors [17]. In the proposed scheme, the optimal relay, which can maximize the minimum residual energy, is selected to cooperate with the source for data transmission. Zhang et al. [18] employ EH-powered spectrum sensors to detect the licensed spectrum to alleviate the spectrum scarcity problem in sensor networks. The authors optimize the detected available time of licensed spectrum while promising the sustainability of spectrum sensors.

2.1.2.2 Prediction-Based Energy Allocation

In prediction-based energy allocation, EH-powered sensor networks obtain the full (causal and non-casual) knowledge of EH processes through prediction. As we have mentioned before, the success of prediction-based energy allocation largely relies on the prediction accuracy. In [19], the authors propose an EH predictor for multiple energy sources (i.e., solar and wind) over horizons ranging from a few minutes to a few hours. The proposed predictor utilizes the correlation in EH process between consecutive time slots, and can achieve 93%–85% accuracy in solar power. In the following, we introduce the prediction-based energy allocations in the literature.

In [2], the authors propose a channel access scheme to maximize the throughput in a point-to-point communication system with the knowledge of energy arrival in K time slots. To deal with the stochasticity in the changes of channel conditions, the authors formulate a discrete-time and continuous-state Markov decision process to schedule channel access and allocate transmission power. Based on the one-step prediction on EH process, Lee et al. propose a cross-layer scheduling to jointly optimize the source rate in transport layer, network throughput in network layer, and duty cycles in media access control (MAC) layer [3].

Numerous works employ mobile sinks to collect data from EH-powered sensors [20–22]. In these scenarios, sensors can transmit data to the sink through one-hop data transmission. Therefore, their energy consumption can be significantly reduced. In [20], Mehrabi et al. maximize the throughput of a sensor network with mobile sinks, in which sensors have constant EH rate. The sink traverses among the sensors to collect data. The authors jointly schedule the speed of the sink and the time slots of data collection. Then [21] extends [20] to a scenario that sensors harvest uniformly distributed energy, and investigates the network throughput maximization problem. In [22], the sink travels in a constant speed along a straight path to collect data from sensors. The authors allocate the time slots for data collection and sensors' data transmission rate to maximize the amount of collected data.

References [23–26] optimize the network utility of large-scale WSNs in which the sensors transmit data to the sink through multi-hop relaying. In [24], Liu et al. design two algorithms to optimize the network utility by exploiting convexity of the network flow problem. The first algorithm computes the data sampling rate and routing based on dual decomposition. To deal with the fluctuations in the EH process, the other algorithm maintains the battery at a target level. In [23], Zhang et al. propose a distributed algorithm to schedule data sensing and perform routing for EHWSNs with limited battery capacity. Furthermore, the proposed algorithm mitigates the estimation error of the EH process by adaptively scheduling the data sensing and routing in each time slot. The authors of [25] present two algorithms for balanced energy allocation of sensors, and optimal data sensing and data transmission. Deng et al. [26] formulate a convex problem to optimize the network utility, taking the coupling of energy neutrality constraints in time and space into consideration. Then, the authors decouple the problem into subproblems by means of dual decomposition and address them by distributed algorithms.

2.1.2.3 Online Energy Allocation

For online energy allocation, sensor networks require only the casual knowledge of the EH process, such as the distribution or the time-average value. In the literature, markov decision process (MDP) and Lyapunov optimization approach have been considered as effective means to design online algorithms.

Numerous works investigate the performance optimization of a point-to-point system with a single transmitter and receiver pair, using MDP-based algorithms [12, 27, 28]. In [27], the authors assume the EH process to be an ergodic stochastic process and design a MDP-based energy management scheme for a sensor with limited battery capacity and data buffer. The proposed scheme can achieve the optimal utility asymptotically while keeping the probabilities of energy outage and data buffer overflow low. In [28], the EH process evolves in an i.i.d. manner. To jointly optimize the transmission delay and energy management, the authors minimize a cost function composed of a convex function of the data buffer length and the energy used for data transmission. A MDP is formulated and addressed by Q-learning algorithm. In [12], the authors consider the imperfect knowledge of the state of charge (SoC) and design a partially observable MDP to optimize the network throughput. The optimization complexity is reduced by decoupling the optimization problem to two time scales: first, optimize the short-term performance w.r.t. the time-varying channel conditions, by enforcing constraints on average energy consumption, and neglecting the battery dynamics; then, optimize the average energy consumption while considering the dynamics in SoC.

In [6, 9, 29, 30], the authors investigate the source estimation using EHWSNs, to improve the estimation accuracy. In [6], the sensors use amplify-and-forward policy to forward the observations. The authors adopt Lyapunov optimization approach to schedule the power amplification factor in an online manner, with the objective to minimize the average mean-square error of source estimation. Knorn et al. [29] investigate a multi-sensor estimation system, in which the sensors not only harvest energy from the ambient energy sources, but also share energy with each other through wireless energy transfer. The authors schedule the data transmission and energy sharing to minimize the distortion. Zordan et al. [9] formulate a constrained MDP to minimize the distortion in data compression while promising the energy buffer level to be larger than a pre-determined threshold. The authors use Lagrangian relaxation approach to transform the problem into an equivalent unconstrained MDP, which can be addressed by a value iteration algorithm. Tapparello et al. [30] jointly study the source coding and data transmission. The authors leverage the correlation of the source measures at closely located sensors to decrease the amount of transmitted data. To deal with the stochasticity in EH process and channel conditions, the authors use Lyapunov optimization to minimize the distortion and ensure the stability of data queues and energy buffers.

The efficient online energy allocation in multi-hop EH-powered sensor networks has also attracted attentions recently [4, 5]. Huang et al. design an online scheduling algorithm which jointly considers the data routing, admission control, and energy management. The algorithm achieves close-to-optimal utility for EHWSNs without a priori knowledge of the EH process [5]. Based on the algorithm in [5], Xu et al. investigate the utility-optimal data sensing and transmission in EHWSNs with heterogeneous energy sources, i.e., power grids and harvested energy, in [4]. Xu et al. also study the trade-off between achieved network utility and cost on energy from power grid.

2.2 Spectrum Harvesting

Although the spectrum harvesting capability alleviates the spectrum scarcity problem in sensor networks, the protection of PUs mandates spectrum sensing before data transmission, which may consume considerable amount of energy of SH sensors. Furthermore, to fully exploit the diversity gain brought by multiple licensed channels, the resource allocation, in terms of transmission channel, power and time, requires careful design to improve both energy and spectrum efficiency. In this section, we overview the related works of spectrum sensing and resource allocation in SHSNs.

2.2.1 Spectrum Sensing

Sensors are mandated to perform spectrum sensing before accessing the licensed spectrum to avoid collisions with PUs. Specially, sensors detect a certain spectrum range to identify the available channels, and then access the channels for data transmission. The popular spectrum-sensing technique includes energy detection, cyclo-stationary detection, and matched filter detection.

Among these techniques, energy detection has been widely considered as a promising solution for spectrum sensing in sensor networks, due to the following reasons: (i) simple implementation; (ii) relatively short sensing time; and (iii) requiring non-priori information of the PU signals. To detect the PU activities, the energy detector accumulates the samples that are overheard from the licensed channels. Then, the energy detector compares the output, i.e., the test statistic of the accumulated samples, to a predefined threshold. The PU is claimed to be active if the output is above the threshold; otherwise, it is claimed to be inactive. In the following, we first survey the works for energy efficient, followed by the EH-powered spectrum-sensing schemes.

2.2.1.1 Energy-Efficient Spectrum Sensing

To accumulate sufficient samples on licensed channel for spectrum sensing, the sensors need to overhear the licensed channel for sample accumulation, which consumes considerable energy and exacerbate the energy scarcity problem in SHSNs [31]. On addressing this issue, a large body of research works propose energy-efficient spectrum-sensing schemes for SHSNs to conserve sensors' scarce energy resource while promising the PU protection.

The basic ideas to achieve energy-efficient spectrum sensing are to decrease the sensing duration [32, 33] and switching times [34], and select optimal detection threshold [35]. Pei et al. [32] investigate energy-efficient wideband spectrum sensing, in which the sensors can detect multiple narrowband channels and aggregate the perceived available channels for data transmission. To decrease the energy consumption in spectrum sensing, the sensing time is an optimized subject to the constraints of detection probability. In [33], the authors consider a two-PU scenario. The sensors allocate sensing time between the two PUs to maximize the aggregate detection probability and find more transmission opportunities. To account for the considerable energy consumption on the frequent on/off switching of sensors, [34] optimizes the schedule order to reduce the switch frequency. In addition to sensing duration and switching times, the energy efficiency of spectrum sensing is also subject to the detection threshold. In [35], the authors minimize the energy consumption of spectrum sensing by optimizing the detection threshold. To this end, a convex optimization problem is formulated and addressed by Lagrangian relaxation.

To exploit the spatial diversity brought by the large number of sensors, numerous researches consider the cooperative spectrum sensing in which multiple sensors simultaneously detect the activity of one PU. In cooperative sensing, the sensors report the sensing results to a fusion center which combines the results to make decision on the PU activity. The reported decisions can be either soft decision, i.e., the test statistics of the accumulated samples, or hard decision, i.e., the one-bit decision on PU activity made by each sensor.

In cooperative spectrum sensing, the possibility of energy conservation in spectrum sensing comes from the observation that some sensors may provide marginal profit in overall performance [36–38]. Deng et al. [36] studied the network lifetime extension of dedicated sensor networks for cooperative spectrum sensing. The authors schedule the on/off of sensors to prolong the network lifetime while promising the detection probability above a threshold. The scheduling of sensors is formulated to a knapsack problem, and then addressed by heuristic algorithms. Li et al. [37] investigate the cooperative spectrum-sensing schedule for a SHSN, in which sensors decide whether to join spectrum sensing for energy conservation. An evolutionary game is formulated to facilitate the decision of sensors according to their utility history. In [38], the authors rank the sensors according to their historic detection accuracy. Then, a selection scheme is selected to find the sensors for cooperative spectrum sensing based on the ranking. Notably, the proposed scheme does not require a priori knowledge of the PU signal-to-noise ratio (SNR).

Several works study the effectiveness of the soft and hard decisions in cooperative spectrum sensing [39–41]. Mu and Tugnait [39] consider soft decision scenario in which the spectrum sensors transmit the continuous-valued sensing test statistics to the fusion center. Based on the sensing statistics, a convex optimization problem is formulated to optimize the SU's sum instantaneous throughput by jointly allocating the transmission power and spectrum access probability. In [40], the authors compare the performance of hard and soft decision-based cooperative sensing schemes with the presence of reporting channel errors. It is shown that hard decision-based sensing is more sensitive to the reporting channel errors, in comparison to the soft decision-based counterpart. Ejaz et al. [41] investigate the spectrum-sensing performance of heterogeneous spectrum harvesting networks, in which the sensors use various spectrum detection techniques for spectrum sensing, including energy detector, cyclostationary detector, etc. The authors compare the performance of these detection techniques with hard and soft decision rules through simulations.

2.2.1.2 EH-Powered Spectrum Sensing

With the emerging of EH technologies, EH-powered spectrum sensing has been gaining more and more attentions to achieve sustainable identification of spectrum opportunities [42–45]. In [42], the authors optimize the throughput of an ESH system, in which sensors cooperatively identify the PU activities. To account for the imprecise information of the PU channel state, a partially observable MDP is formulated to obtain the optimal cooperation among sensors. Zhang et al. [44] consider a scenario consisting of one PU and one sensor which both harvest energy from the same ambient energy source. By exploiting the correlation of harvested energy at the two nodes, the authors propose a two-dimensional spectrum and power-sensing scheme to improve the PU detection performance. Nobar et al. [43] analyze the performance of a RF-powered SH network which consists of a central node with EH capability and multiple sensors. The central node shares the harvested energy with the sensors through RF wireless charging. The authors analyze the throughput of sensors with fixed energy consumption on spectrum sensing. In [45], the authors consider a dual-hop SH network which incorporates multiple amplify-and-forward relays. The relays cooperatively detect the presence of PUs. According to the sensing decisions on the PU activity, the relays switch between data transmission and RF energy harvesting modes to transmit data or recharge the batteries, respectively. The authors study the system performance in terms of detection probability, average harvested energy, and outage probability.

2.2.2 Resource Allocation in Spectrum Harvesting Sensor Networks

Through spectrum sensing, sensors obtain the knowledge of licensed spectrum availability to facilitate data transmission. To fully benefit from the multi-channel diversity provided by SH capability, one needs to carefully allocate sensors' transmission power and access time over the available channels. Furthermore, the utilization of licensed spectrum should take the possibility of returning PU into consideration, and timely vacate the channel to avoid collisions. In the following, we overview the related works regarding protocol design and dynamic spectrum access in SHSNs.

2.2.2.1 Protocol Design

Numerous MAC protocols have been proposed for efficient spectrum access in SHSNs, based on carrier sense multiple access with collision avoidance (CSMA/CA) [46–48], and time division multiple access (TDMA) [49]. Tan and Le [46] propose a synchronized MAC, in which time is divided into fixed-size cycles consisting of three phases, i.e., the sensing phase, the synchronization phase, and the data transmission phase. The sensors detect the presence of PUs in the sensing phase and broadcast beacon signals for synchronization. In the data transmission phase, the sensors perform contention using CSMA/CA protocol to access the channels. Based on the protocol, the authors optimize the sensing period and contention window to maximize the network throughput, subject to the detection probability constraints for PUs. In [47], the authors extended the traditional RTS/CTS handshake process to PTS/RTS/CTS process. Once a sensor has data packet to transmit and detect a idle channel, it sends a Prepare-To-Sense message to neighbors and asks them to keep silence in the following duration. If the PU is detected to be present, the sensors suspend their backoff timer for a blocking time. The authors analyze the performance of the proposed protocol in terms of throughput and average packet service time. Shah and Akan [48] propose a CSMA-based cognitive adaptive MAC (CAMAC) which decrease the energy consumption on spectrum sensing. CAMAC exploits the spatial correlation of densely deployed sensors and only use a small number of sensors to sense licensed channels. The outcomes of spectrum sensing are shared with the nearby sensors for data transmission. Anamalamudi and Jin [49] design a hybrid common control channel (CCC)-based CR-MAC protocol, in which the sensors exchange control information through CCC in a TDMA manner and compete to access the available licensed channels through CSMA/CA.

In addition, [50–52] focus on the routing protocol design for multi-hop SHSNs. Ozger et al. [50] design a cluster-based routing protocol, in which sensors form clusters to deliver information through multi-hop relaying. Each cluster has at least one common channel for data transmission between the members and the head. The sensors with larger eligible node degree, more number of available channels, and more remaining energy are likely to be selected as the cluster heads. Considering the spectrum mobility of PUs, the authors analyze the average re-clustering probability. The routing protocol proposed in [51] exploits dedicated cluster heads with infinite energy supply to organize the sensors into clusters. Spachos and Hantzinakos [52] design a reactive routing protocol, in which the destination sensor initiates the route discovery process. The sensor that is closer to the destination has higher priority to relay data packets.

References [53, 54] consider cross-layer protocols for SHSNs to improve the network performance. Ping et al. [53] propose a spectrum aggregation-based cross-layer protocol. In physical layer and MAC layer, the sensors aggregate the available channels for data transmission. In the routing layer, the authors consider three different routing selection criterions, i.e., energy efficiency, throughput maximization, and delay minimization. Shah et al. [54] propose a QoS-aware cross-layer protocol for SHSNs in smart grid applications. The authors exploit SH capability to mitigate the noisy and congested channels. The authors jointly allocate channels, schedule flow rate, and decide routings to meet the diverse QoS requirements of smart grid applications, by exploiting Lyapunov optimization approach.

2.2.2.2   Dynamics Spectrum Access

To account for highly dynamic PU activities, numerous works model the PU activity by stochastic processes and design spectrum access schemes for SHSNs [31, 55–57]. In [55] and [56], the authors model the PU activities by Markov process and take the spectrum detection errors into consideration. Urgaonkar and Neely [55] develop an opportunistic channel accessing policy to maximize the network throughput subject to the maximum collision constraint. In [56], Qin et al. optimize the delay and throughput of a multi-hop SHSN in which sensors are mounted with multiple CR transceivers. Liang et al. [31] investigate the sensing-throughput trade-off in SH systems which jointly realize spectrum sensing and access. The authors optimize the sensing duration to maximize the throughput while guaranteeing the PU protection. Sharma and Sahoo [57] model the channel occupancy as a renewal process. Sensors randomly detect the licensed channel to decrease the spectrum sensing overhead. The authors analyze the accessible time of idle channels subject to the collision probability with PUs.

There are also a large body of works that exploit game theory to design distributed spectrum access for SHSNs [58–62]. Zhang et al. [58] design a multi-channel access scheme by congestion game, in which the SUs strive to access the channel to maximize their own utility. Zheng et al. [59] consider a canonical scenario in which the users are stochastically active due to their data service requirement. The authors model interferences between users by a dynamic interference graph, based on which a graphical stochastic game is formulated to schedule the channel access and mitigate the interferences. By proving the game to be an exact potential game, the existence of the Nash equilibrium can be guaranteed. Xu et al. [60] model the PU activities by Bernoulli processes. To enable distributed channel selection, the authors formulate the channel selection problem as an exact potential game. Considering the stochasticity in PU activities, a stochastic learning automata-based channel selection algorithm is proposed, with which sensors learn from their individual action–reward history and adjust their behaviors toward a Nash equilibrium point. Rawat et al. [61] consider a scenario in which sensors purchase licensed spectrum access opportunities from the spectrum provider (SP). The authors design a two-stage Stahckelberg game with the SP as the leader and the sensors as the follower. In the first stage, the sensors maximize their payoffs while considering the constraints on transmission power and budget. In the second stage, the SPs offer competitive prices for spectrum usage to maximize their revenues subject to their system capabilities. In [62], the authors propose a two-layered game for a scenario in which the PUs share the spectrum resource with SUs to gain revenue. A two-layered game is formulated for revenue sharing between the operators of a primary network and a secondary network. The top layer forms a Nash bargaining game to determine the revenue sharing scheme, while the bottom layer forms a Stackelberg game to achieve optimal resource allocation.

2.3 Joint Energy and Spectrum Harvesting in Wireless Networks

Nowadays, the exponential growth of wireless devices leads to a continuous surge in both energy consumption and network capacity. To mitigate the caused energy and spectrum scarcity issues, it is desirable to enhance the wireless devices with ESH capabilities. In the literature, the existing works related to ESH wireless networks can be divided into two categories: energy harvested from environmental energy sources (referred to as green energy hereafter) and energy harvested from the RF signal of the PUs.

2.3.1 Green Energy-Powered SH Networks

References [63, 64] focus on the joint schedule of spectrum sensing and access powered by green energy, in which the EH process is independent of PU operation. In [63], Park et al. investigate the throughput maximization by modeling the PU activity as a discrete Markov process and the EH process as i.i.d. process. The authors formulate the stochastic optimization problem as a partially observable MDP, based on which a spectrum-sensing policy and an optimal detection threshold are jointly designed. Using the similar modeling of EH process and PU activities in [63, 64] finds that the operation of green energy-powered SH networks can be characterized in terms of a spectrum-limited regime and an energy-limited regime, depending on the detection threshold. In the former regime, the system has sufficient energy for spectrum access, while in the latter regime, the availability of energy resource limits the number of spectrum access attempts. The authors analyze the probability to access an idle channel and an occupied channel, based on which they derive an optimal detection threshold to maximize the throughput.

References [22, 65] investigate the energy and spectrum allocation in large-scale ESHSNs. Ren et al. [22] optimize the spectrum access and energy allocation to optimize the network utility. However, [22] requires non-causal information of EH process and PU activities. In [65], the authors develop an aggregate network utility optimization framework for the design of an online energy and spectrum management based on Lyapunov optimization. The framework captures three stochastic processes, i.e., the EH process, PU activities, and channel conditions.

2.3.2 RF-Powered SH Networks

References [66–68] investigate the spectrum access and energy allocation of RF-powered SHSNs, in which the sensors harvest energy from the PU signal. Hoang et al. [66] consider a network where the sensor can perform channel access when the selected channel is idle, or harvest energy when the channel is occupied by PUs. The authors formulate an optimization formulation based on MDP to obtain channel access policy, with the objective to maximize the throughput. Furthermore, an online learning algorithm is employed to obtain the system parameter. Hoang et al. [67] extend the scenario in [66] to a cooperative multi-user network. To deal with the high-complexity brought by the MDP-based algorithm, the authors use a dual decomposition method to distributedly solve the stochastic problem. Lee et al. [68] use a stochastic-geometry model in which the sensors and PUs are randomly deployed in the AoI. Based on the stochastic-geometry model, [68] analyzes the transmission probability and the spatial throughput of sensors. In addition, the authors find the optimal transmission power and density of sensors to optimize the secondary network throughput.

2.4 Conclusion

This chapter provides a literature survey of resource allocation related to ESHSNs. First, we overview the existing works on EH process modeling and allocation in EHSNs. Then, the spectrum sensing and allocation in SHSNs are discussed. At last, we introduce the works which jointly consider energy and spectrum allocation in wireless networks with ESH capabilities. In the subsequent chapters, the EH-powered spectrum sensing and the joint allocation of energy and spectrum in ESHSNs will be studied.

References

1. M.L. Ku, W. Li, Y. Chen, K.J.R. Liu, Advances in energy harvesting communications: past, present, and future challenges. IEEE Commun. Surv. Tutorials **18**(2), 1384–1412 (2016)
2. Z. Wang, V. Aggarwal, X. Wang, Power allocation for energy harvesting transmitter with causal information. IEEE Trans. Commun. **62**(11), 4080–4093 (2014)
3. S. Lee, B. Kwon, S. Lee, A.C. Bovik, Bucket: scheduling of solar-powered sensor networks via cross-layer optimization. IEEE Sensors J. **15**(3), 1489–1503 (2015)
4. W. Xu, Y. Zhang, Q. Shi, and X. Wang, Energy management and cross layer optimization for wireless sensor network powered by heterogeneous energy sources, IEEE Trans. Wirel. Commun. **14**(5), 2814–2826 (2015)
5. L. Huang, M. Neely, Utility optimal scheduling in energy-harvesting networks. IEEE/ACM Trans. Netw. **21**(4), 1117–1130 (2013)
6. H. Zhou, N. Cheng, N. Lu, L. Gui, D. Zhang, Q. Yu, F. Bai, X.S. Shen, Whitefi infostation: engineering vehicular media streaming with geolocation database. IEEE J. Sel. Areas Commun. **34**(8), 2260–2274 (2016)
7. W. Liu, X. Zhou, S. Durrani, H. Mehrpouyan, S.D. Blostein, Energy harvesting wireless sensor networks: delay analysis considering energy costs of sensing and transmission. IEEE Trans. Wirel. Commun. **15**(7), 4635–4650 (2016)
8. C. Yang, K.W. Chin, On nodes placement in energy harvesting wireless sensor networks for coverage and connectivity. IEEE Trans. Ind. Inf. (to be published)
9. D. Zordan, T. Melodia, M. Rossi, On the design of temporal compression strategies for energy harvesting sensor networks. IEEE Trans. Wirel. Commun. **15**(2), 1336–1352 (2016)
10. D. Niyato, E. Hossain, A. Fallahi, Sleep and wakeup strategies in solar-powered wireless sensor/mesh networks: performance analysis and optimization. IEEE Trans. Mobile Comput. **6**(2), 221–236 (2007)
11. N. Michelusi, M. Zorzi, Optimal adaptive random multiaccess in energy harvesting wireless sensor networks. IEEE Trans. Commun. **63**(4), 1355–1372 (2015)
12. N. Michelusi, L. Badia, M. Zorzi, Optimal transmission policies for energy harvesting devices with limited state-of-charge knowledge. IEEE Trans. Commun. **62**(11), 3969–3982 (2014)
13. D.D. Testa, N. Michelusi, M. Zorzi, Optimal transmission policies for two-user energy harvesting device networks with limited state-of-charge knowledge. IEEE Trans. Wirel. Commun. **15**(2), 1393–1405 (2016)
14. M.L. Ku, Y. Chen, K.J.R. Liu, Data-driven stochastic models and policies for energy harvesting sensor communications. IEEE J. Sel. Areas Commun. **33**(8), 1505–1520 (2015)
15. E. Ibarra, A. Antonopoulos, E. Kartsakli, J.J.P.C. Rodrigues, C. Verikoukis, QoS-aware energy management in body sensor nodes powered by human energy harvesting. IEEE Sensors J. **16**(2), 542–549 (2016)
16. A. Kansal, J. Hsu, S. Zahedi, M.B. Srivastava, Power management in energy harvesting sensor networks. ACM Trans. Embed. Comput. Syst. **6**(4) (2007). doi:10.1145/1274858.1274870
17. D. Zhang, Z. Chen, H. Zhou, L. Chen, X. Shen, Energy-balanced cooperative transmission based on relay selection and power control in energy harvesting wireless sensor network. Comput. Netw. **104**, 189–197 (2016)
18. D. Zhang, Z. Chen, J. Ren, Z. Ning, K.M. Awad, H. Zhou, X. Shen, Energy harvesting-aided spectrum sensing and data transmission in heterogeneous cognitive radio sensor network. IEEE Trans. Veh. Technol. (to be published). doi:10.1109/TVT.2016.2551721
19. A. Cammarano, C. Petrioli, D. Spenza, Online energy harvesting prediction in environmentally powered wireless sensor networks. IEEE Sensors J. **16**(17), 6793–6804 (2016)
20. A. Mehrabi, K. Kim, General framework for network throughput maximization in sink-based energy harvesting wireless sensor networks. IEEE Trans. Mobile Comput. (to be published). doi:10.1109/TMC.2016.2607716
21. A. Mehrabi, K. Kim, Maximizing data collection throughput on a path in energy harvesting sensor networks using a mobile sink. IEEE Trans. Mobile Comput. **15**(3), 690–704 (2016)

22. J. Ren, Y. Zhang, R. Deng, N. Zhang, D. Zhang, X. Shen, Joint channel access and sampling rate control in energy harvesting cognitive radio sensor networks. IEEE Trans. Emerg. Top. Comput. (to be published). doi:10.1109/TETC.2016.2555806
23. Y. Zhang, S. He, J. Chen, Y. Sun, X. Shen, Distributed sampling rate control for rechargeable sensor nodes with limited battery capacity. IEEE Trans. Wirel. Commun. 12(6), 3096–3106 (2013)
24. R.-S. Liu, P. Sinha, C. Koksal, Joint energy management and resource allocation in rechargeable sensor networks, in *Proceedings of IEEE INFOCOM* (2010)
25. Y. Zhang, S. He, J. Chen, Data gathering optimization by dynamic sensing and routing in rechargeable sensor networks. IEEE/ACM Trans. Netw. 24(3), 1632–1646 (2016)
26. R. Deng, Y. Zhang, S. He, J. Chen, X. Shen, Maximizing network utility of rechargeable sensor networks with spatiotemporally coupled constraints. IEEE J. Sel. Areas Commun. 34(5), 1307–1319 (2016)
27. R. Srivastava, C.E. Koksal, Basic performance limits and tradeoffs in energy-harvesting sensor nodes with finite data and energy storage. IEEE/ACM Trans. Netw. 21(4), 1049–1062 (2013)
28. K.J. Prabuchandran, S.K. Meena, S. Bhatnagar, Q-learning based energy management policies for a single sensor node with finite buffer. IEEE Wirel. Commun. Lett. 2(1), 82–85 (2013)
29. S. Knorn, S. Dey, A. Ahln, D.E. Quevedo, Distortion minimization in multi-sensor estimation using energy harvesting and energy sharing. IEEE Trans. Signal Process. 63(11), 2848–2863 (2015)
30. C. Tapparello, O. Simeone, M. Rossi, Dynamic compression-transmission for energy-harvesting multihop networks with correlated sources. IEEE/ACM Trans. Netw. 22(6), 1729–1741 (2014)
31. Y.-C. Liang, Y. Zeng, E. Peh, A.T. Hoang, Sensing-throughput tradeoff for cognitive radio networks. IEEE Trans. Wirel. Commun. 7(4), 1326–1337 (2008)
32. Y. Pei, Y.C. Liang, K.C. Teh, K.H. Li, How much time is needed for wideband spectrum sensing? IEEE Trans. Wirel. Commun. 8(11), 5466–5471 (2009)
33. I. Kim, D. Kim, Optimal allocation of sensing time between two primary channels in cognitive radio networks. IEEE Commun. Lett. 14(4), 297–299 (2010)
34. P. Cheng, R. Deng, J. Chen, Energy-efficient cooperative spectrum sensing in sensor-aided cognitive radio networks. IEEE Wirel. Commun. 19(6), 100–105 (2012)
35. A. Ebrahimzadeh, M. Najimi, S. Andargoli, A. Fallahi, Sensor selection and optimal energy detection threshold for efficient cooperative spectrum sensing. IEEE Trans. Veh. Technol. 64(4), 1565–1577 (2015)
36. R. Deng, J. Chen, C. Yuen, P. Cheng, Y. Sun, Energy-efficient cooperative spectrum sensing by optimal scheduling in sensor-aided cognitive radio networks. IEEE Trans. Veh. Technol. 61(2), 716–725 (2012)
37. H. Li, X. Xing, J. Zhu, X. Cheng, K. Li, R. Bie, T. Jing, Utility-based cooperative spectrum sensing scheduling in cognitive radio networks. IEEE Trans. Veh. Technol. (to be published). doi:10.1109/TVT.2016.2532886
38. Z. Khan, J. Lehtomaki, K. Umebayashi, J. Vartiainen, On the selection of the best detection performance sensors for cognitive radio networks. IEEE Signal Process. Lett. 17(4), 359–362 (2010)
39. H. Mu, J.K. Tugnait, Joint soft-decision cooperative spectrum sensing and power control in multiband cognitive radios. IEEE Trans. Signal Process. 60(10), 5334–5346 (2012)
40. S. Chaudhari, J. Lunden, V. Koivunen, H.V. Poor, Cooperative sensing with imperfect reporting channels: Hard decisions or soft decisions? IEEE Trans. Signal Process. 60(1), 18–28 (2012)
41. W. Ejaz, G. Hattab, N. Cherif, M. Ibnkahla, F. Abdelkefi, M. Siala, Cooperative spectrum sensing with heterogeneous devices: hard combining versus soft combining. IEEE Syst. J. (99), 1–12 (2016)
42. P. Pratibha, K.H. Li, K.C. Teh, Dynamic cooperative sensing-access policy for energy-harvesting cognitive radio systems. IEEE Trans. Veh. Technol. (to be published). doi:10.1109/TVT.2016.2532900

43. S.K. Nobar, K.A. Mehr, J.M. Niya, RF-powered green cognitive radio networks: architecture and performance analysis. IEEE Commun. Lett. **20**(2), 296–299 (2016)
44. W. Zhang, Y. Guo, H. Liu, Y. Chen, Z. Wang, J. Mitola, Distributed consensus-based weight design for cooperative spectrum sensing. IEEE Trans. Parallel Distrib. Syst. **26**(1), 54–64 (2015)
45. N.I. Miridakis, T.A. Tsiftsis, G.C. Alexandropoulos, M. Debbah, Green cognitive relaying: opportunistically switching between data transmission and energy harvesting. IEEE J. Sel. Areas Commun. (to be published)
46. L.T. Tan, L.B. Le, Distributed MAC protocol for cognitive radio networks: design, analysis, and optimization. IEEE Trans. Veh. Technol. **60**(8), 3990–4003 (2011)
47. Q. Chen, W.C. Wong, M. Motani, Y.C. Liang, MAC protocol design and performance analysis for random access cognitive radio networks. IEEE J. Sel. Areas Commun. **31**(11), 2289–2300 (2013)
48. G. Shah, O. Akan, Cognitive adaptive medium access control in cognitive radio sensor networks. IEEE Trans. Veh. Technol. **64**(2), 757–767 (2015)
49. S. Anamalamudi, M. Jin, Energy-efficient hybrid CCC-based MAC protocol for cognitive radio ad hoc networks. IEEE Syst. J. **10**(1), 358–369 (2016)
50. M. Ozger, E. Fadel, O.B. Akan, Event-to-sink spectrum-aware clustering in mobile cognitive radio sensor networks. IEEE Trans. Mobile Comput. **15**(9), 2221–2233 (2016)
51. P.T.A. Quang, D.S. Kim, Throughput-aware routing for industrial sensor networks: application to ISA100.11a. IEEE Trans. Ind. Inf. **10**(1), 351–363 (2014)
52. P. Spachos, D. Hantzinakos, Scalable dynamic routing protocol for cognitive radio sensor networks. IEEE Sensors J. **14**(7), 2257–2266 (2014)
53. S. Ping, A. Aijaz, O. Holland, A.H. Aghvami, SACRP: a spectrum aggregation-based cooperative routing protocol for cognitive radio ad-hoc networks. IEEE Trans. Commun. **63**(6), 2015–2030 (2015)
54. G.A. Shah, V.C. Gungor, O.B. Akan, A cross-layer QoS-aware communication framework in cognitive radio sensor networks for smart grid applications. IEEE Trans. Ind. Inf. **9**(3), 1477–1485 (2013)
55. R. Urgaonkar, M. Neely, Opportunistic scheduling with reliability guarantees in cognitive radio networks. IEEE Trans. Mobile Comput. **8**(6), 766–777 (2009)
56. Y. Qin, J. Zheng, X. Wang, H. Luo, H. Yu, X. Tian, X. Gan, Opportunistic scheduling and channel allocation in MC-MR cognitive radio networks. IEEE Trans. Veh. Technol. **63**(7), 3351–3368 (2014)
57. M. Sharma, A. Sahoo, Stochastic model based opportunistic channel access in dynamic spectrum access networks. IEEE Trans. Mobile Comput. **13**(7), 1625–1639 (2014)
58. N. Zhang, H. Liang, N. Cheng, Y. Tang, J. Mark, X. Shen, Dynamic spectrum access in multi-channel cognitive radio networks. IEEE J. Sel. Areas Commun. **32**(11), 2053–2064 (2014)
59. J. Zheng, Y. Cai, N. Lu, Y. Xu, X. Shen, Stochastic game-theoretic spectrum access in distributed and dynamic environment. IEEE Trans. Veh. Technol. **64**(10), 4807–4820 (2015)
60. Y. Xu, J. Wang, Q. Wu, A. Anpalagan, Y.D. Yao, Opportunistic spectrum access in unknown dynamic environment: a game-theoretic stochastic learning solution. IEEE Trans. Wirel. Commun. **11**(4), 1380–1391 (2012)
61. D.B. Rawat, S. Shetty, C. Xin, Stackelberg-game-based dynamic spectrum access in heterogeneous wireless systems. IEEE Syst. J. (to be published). doi:10.1109/JSYST.2014.2347048
62. Y. Wu, Q. Zhu, J. Huang, D.H.K. Tsang, Revenue sharing based resource allocation for dynamic spectrum access networks. IEEE J. Sel. Areas Commun. **32**(11), 2280–2296 (2014)
63. S. Park, D. Hong, Optimal spectrum access for energy harvesting cognitive radio networks. IEEE Trans. Wirel. Commun. **12**(12), 6166–6179 (2013)
64. S. Park, H. Kim, D. Hong, Cognitive radio networks with energy harvesting. IEEE Trans. Wirel. Commun. **12**(3), 1386–1397 (2013)
65. D. Zhang, Z. Chen, M.K. Awad, N. Zhang, H. Zhou, X.S. Shen, Utility-optimal resource management and allocation algorithm for energy harvesting cognitive radio sensor networks. IEEE J. Sel. Areas Commun. **34**(12), 3552–3565 (2016). doi:10.1109/JSAC.2016.2611960

66. D.T. Hoang, D. Niyato, P. Wang, D.I. Kim, Opportunistic channel access and RF energy harvesting in cognitive radio networks. IEEE J. Sel. Areas Commun. 32(11), 2039–2052 (2014)
67. D.T. Hoang, D. Niyato, P. Wang, D.I. Kim, Performance optimization for cooperative multiuser cognitive radio networks with RF energy harvesting capability. IEEE Trans. Wirel. Commun. 14(7), 3614–3629 (2015)
68. S. Lee, R. Zhang, K. Huang, Opportunistic wireless energy harvesting in cognitive radio networks. IEEE Trans. Wirel. Commun. 12(9), 4788–4799 (2013)

Chapter 3
Spectrum Sensing and Access in Heterogeneous SHSNs

In this chapter, we investigate the spectrum sensing and access of a heterogeneous spectrum harvesting sensor network (HSHSN) which consists of EH-enabled spectrum sensors and battery-powered data sensors. The former detects the availability of licensed channels, while the latter transmits sensed data to the sink over the available channels. Two algorithms that operate in tandem are proposed to achieve the sustainability of spectrum sensors and conserve energy of data sensors, while the EH dynamics and PU protections are considered. Extensive simulation results are given to validate the effectiveness of the proposed algorithms.

3.1 Introduction

To avoid interfering with the PUs, sensors perform spectrum sensing to identify PU activities before transmitting data. Considering the fact that single-node spectrum sensing suffers from the spatially large-scale effect of shadowing and small-scale effect of multi-path fading, cooperative spectrum sensing is preferred to enhance the spectrum accuracy. However, as mentioned in Chap. 2, the cooperative spectrum sensing consumes considerable amount of energy which may deteriorate SHSNs' lifetime. On addressing this issue, energy harvesting (EH) has been considered as a promising solution to recharge the batteries by converting the renewable energy to electric power [1]. Using harvested energy, the spectrum sensors can sustainably identify PU activities without battery replacement. In the literature, a few works investigate the scheduling of renewable energy-powered spectrum sensing [2, 3]. However, they either consider a single-spectrum sensor scenario [3], or design algorithms based on Markov decision process (MDP) which may not be scalable in large-scale sensor networks [2]. In addition, the radio frequency (RF)-powered spectrum sensing has been investigated in [4, 5] which require dedicated energy sources to provide wireless energy charging, and thus increase the deployment cost of SHSNs.

© The Author(s) 2017
D. Zhang et al., *Resource Management for Energy and Spectrum Harvesting Sensor Networks*, SpringerBriefs in Electrical and Computer Engineering, DOI 10.1007/978-3-319-53771-9_3

Based on the licensed channel availability, sensors are allowed to access the available ones for data transmission. Comparing to the fixed channel allocation in traditional WSNs, multiple accessible channels introduce new possibility to improve the energy efficiency in data transmission. The existing works consider either spectrum allocation [6, 7] or power control [8, 9]. However, since sensors' energy consumption is determined by their transmission time and power over the spectrum, limitation remains for the existing solutions which treat the spectrum or power allocation separately.

On addressing the above issues, we make an effort to investigate the green energy-powered spectrum sensing and energy-efficient spectrum access. We consider a specific networking paradigm of ESHSNs, i.e., a heterogeneous SHSN (referred to as HSHSN hereafter) which consists of EH-powered spectrum sensors and battery-powered data sensors. The HSHSN operates over two phases, i.e., a spectrum-sensing phase followed by a data transmission phase. In the former phase, EH-enabled spectrum sensors cooperatively sense the spectrum to discover available licensed channels. Spectrum-sensing scheduling is optimized to maximize the detected channel's available time considering the dynamics of EH. In the latter phase, the data sensors access the available channels to transmit the sensed data. The spectrum and access decisions in each phase are combined in a unified solution to jointly guarantee the accuracy of spectrum sensing, the sustainability of spectrum sensors, and the energy efficiency of the data sensors. Notably, both the algorithms designed for the two phases are computationally efficient and scalable in large-scale SHSNs.

This chapter is organized as follows: the network architecture and spectrum-sensing model are detailed in Sect. 3.2. A mathematical formulation problem and the proposed solutions for the spectrum sensor scheduling problem and data sensors resource allocation problem are detailed in Sect. 3.3. Performance evaluation results are provided in Sect. 3.4. Section 3.5 concludes this chapter.

3.2 System Model

In this section, we describe the network architecture of the heterogeneous spectrum harvesting sensor network (HSHSN), and the EH-powered spectrum-sensing model.

3.2.1 Network Architecture

The HSHSN under consideration coexists with a primary network that has privilege to access the licensed spectrum. The licensed spectrum is divided into K orthogonal channels with equal bandwidth W. The HSHSN consists of three types of nodes: N battery-powered data sensors, M EH-enabled spectrum sensors, and a sink node, as shown in Fig. 3.1. Spectrum sensors are responsible to identify available channels that are not utilized by PUs, whereas data sensors are responsible to collect data

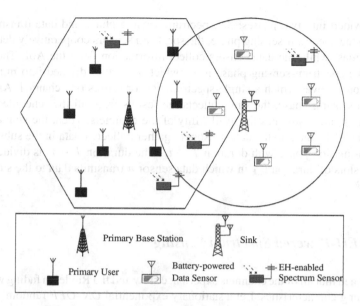

Fig. 3.1 An illustration of the heterogeneous spectrum harvesting sensor network

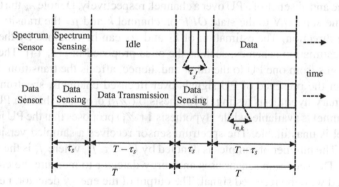

Fig. 3.2 Timing diagram and frame structure of the HSHSN

from an AoI. The sink gathers the data from the data sensors through the available channels.

The operation of the HSHSN is as follows: First, the sink assigns licensed channels to spectrum sensors for PU activity detection, using energy detection [10]. One channel is determined to be unavailable, i.e., PU is active, if at least one scheduled spectrum sensor reports its presence [11]. The power consumption of spectrum sensing is denoted by P_s. The EH rate is assumed to be a priori and keeps stable over T [12]. The EH rate of spectrum sensor m is denoted π_m. After the spectrum sensing, the available channels are allocated to the data sensors for data transmission.

Figure 3.2 shows the timing diagram and frame structure of the considered HSHSN. The HSHSN operates periodically over time slots of duration T. One time slot is divided into two phases: the spectrum-sensing phase and data transmission phase. In the spectrum-sensing phase, the spectrum sensors cooperatively detect the PU activities, while the data sensors collect information from the AoI. The duration of the spectrum-sensing phase is τ_s, which is further divided into mini-slots of duration $\tau_{s'}$ over which a single-spectrum sensor senses one channel. After the spectrum-sensing phase, the sink collects the results from all the scheduled spectrum sensors and estimates the availability of the channels. Then, the sink assigns the available channels to the data sensors to gather collected data in the subsequent data transmission phase with duration $T - \tau_s$. The duration $T - \tau_s$ is divided over the time slots of duration $t_{n,k}$ in which data sensor n transmits data to the sink over channel k.

3.2.2 EH-Powered Spectrum Sensing

It is assumed that all of the channels experience slow and flat Rayleigh fading with the same fading characteristics. Let a stationary exponential ON–OFF random process model the PU behavior over each channel, in which the ON and OFF states represent the presence and absence of a PU over a channel, respectively. Denote λ_k the transition rate from the state ON to the state OFF on channel k and μ_k the transition rate in the reverse direction. The estimation of λ_k and μ_k can be obtained by the channel parameter estimation schemes, such as the ones proposed in [13, 14]. The channel usage changes from one PU to the other and, hence, affects the transition rates.

To detect the presence of PU signals over licensed channels, spectrum sensors perform binary hypothesis testing. Hypothesis 0 ($\mathcal{H}_0$) proposes that the PU is OFF and the channel is available, while Hypothesis 1 ($\mathcal{H}_1$) proposes that the PU is ON and the channel is unavailable. The spectrum sensor receives a sampled version of the PU signal. The number of samples is denoted by $U = \tau_s f_s$, where f_s is the sampling frequency. The spectrum sensor uses an energy detector to measure the energy that is associated with the received signal. The output of the energy detector, i.e., the test statistic, is compared to the detection threshold ε, to make a decision on the state of the PU, ON or OFF. The test statistic evaluates to $Y_{m,k} = \frac{1}{U} \sum_{u=1}^{U} |y_{m,k}(u)|^2$, where $y_{m,k}(u)$ is the u-th sample of the received signal at spectrum sensor m on channel k. The PU signal is a complex-valued PSK signal and the noise is circularly symmetric complex Gaussian with zero mean and σ^2 variance [15].

The performance of the energy detector is evaluated by the following metrics under hypothesis testing [16]:

- The false alarm probability $p_f(m, k)$: The probability that spectrum sensor m detects a PU to be present on channel k when it is not present in fact, i.e., $\mathcal{H}_0$ is true. The false alarm probability is given by Liang et al. [15]

$$p_f(m, k) = Pr(Y_{m,k} > \varepsilon|\mathcal{H}_0) = Q\left(\left(\frac{\varepsilon}{\sigma^2} - 1\right)\sqrt{U}\right), \qquad (3.1)$$

where $Q(\cdot)$ is the complementary distribution function of the standard Gaussian. Without loss of generality, the detection threshold is set to be the same for all of the spectrum sensors; hence, the false alarm probability becomes fixed for all of the spectrum sensors and is denoted by $\bar{p}_f$.

- The detection probability $p_d(m, k)$: The probability that spectrum sensor m detects the presence of a PU on channel k when it is present in fact, i.e., $\mathcal{H}_1$ is true. This probability is given by Liang et al. [15]

$$p_d(m, k) = Pr(Y_{m,k} > \varepsilon | \mathcal{H}_1) = Q\left(\frac{Q^{-1}(\bar{p}_f) - \sqrt{U} \gamma_{m,k}}{\sqrt{2\gamma_{m,k} + 1}} \right), \qquad (3.2)$$

where $\gamma_{m,k}$ denotes the received signal-to-noise ratio (SNR) of spectrum sensor m from the PU on channel k. To reduce the communication overhead and delay, each spectrum sensor sends the final 1-bit decision (e.g., 0 or 1 represents the *ON* or *OFF* state, respectively) to the sink. The final decision on the presence of a PU is made following the logic OR rule [11]. Under this rule, the sink determines the PU to be present if at least one of the scheduled sensors reports that its presence. Therefore, we can express the final false alarm probability F_f^k and final detection probability F_d^k as

$$F_f^k = 1 - \Pi_{m \in \mathcal{M}_k}(1 - \bar{p}_f), \text{ and} \qquad (3.3)$$

$$F_d^k = 1 - \Pi_{m \in \mathcal{M}_k}(1 - p_d(m, k)), \qquad (3.4)$$

where $\mathcal{M}_k$ denotes the set of spectrum sensors scheduled to detect channel k.

3.3 Problem Statement and Proposed Solution

Considering the above-described architecture of the HSHSN and EH dynamics, the scheduling of the spectrum sensors and data sensors becomes challenging. In the spectrum sensor scheduling (SSS) problem, the sink schedules the spectrum sensors to sense the presence of the PUs with the objective to maximize the detected available channels, taking into consideration the EH dynamics and PUs' priorities in accessing the channels. Solving this problem reveals the available channels to the sink which allocates them to the battery-powered data sensors along with the transmission time and power in such a way that the data sensors' energy consumption is minimized. We refer this resource allocation problem as the data sensor resource allocation (DSRA) problem.

Figure 3.3 shows the diagram of the two problems, and the data flows among them. The following two subsections present problem formulations and solutions for both problems, respectively. We formulate the first problem as a nonlinear integer programming problem, and the second problem as a biconvex optimization problem.

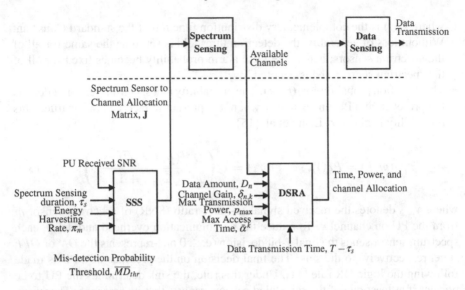

Fig. 3.3 A block diagram of the proposed system. The *dashed line* separates the optimization plane from the sensing plane

3.3.1 Spectrum-Sensing Scheduling

This subsection investigates the SSS problem that is posed as a nonlinear integer programming problem. The problem is solved by a cross-entropy-based solution to maximize the detected available time of channels, while guaranteeing the sustainability of EH-powered spectrum sensors and PUs' protection.

3.3.1.1 Problem Formulation

For the spectrum sensors, three factors impact the average detected available time of the channel: its actual average available time, the final false alarm probability F_f^k, and the final detection probability F_d^k. The actual average available time of channel k evaluates to the product of the mean sojourn time and the stationary probability of channel k. Denote $\bar{L}_{ON}^k = \frac{1}{\lambda_k}$ and $\bar{L}_{OFF}^k = \frac{1}{\mu_k}$ the mean sojourn time of the ON state and the OF state on channel k, respectively. Moreover, the stationary probabilities of the ON and OFF states are given by

$$P_{ON}^k = \frac{\mu_k}{\lambda_k + \mu_k}, \quad P_{OFF}^k = \frac{\lambda_k}{\lambda_k + \mu_k}. \tag{3.5}$$

Therefore, the average available time of channel k can be given by

$$\alpha^k = \bar{L}_{OFF}^k \cdot P_{ON}^k. \tag{3.6}$$

Let **J** be an $M \times K$ matrix with binary elements $J_{m,k}$. $J_{m,k} = 1$ indicates the assignment of spectrum sensor m to detect channel k and 0 otherwise. Given that the PU on channel k is off, the probability that channel k is detected to be available equals to the complement of the final false alarm probability, which can be written as

$$1 - F_f^k = \Pi_{m \in \mathcal{M}_k}(1 - \bar{p}_f) = (1 - \bar{p}_f)^{\sum_{m=1}^{M} J_{m,k}}. \tag{3.7}$$

The data sensor transmission interferes with that of the PUs if the spectrum sensors do not detect the PU presence while it is active. The chance of this event is captured by the misdetection probability $1 - F_d^k$. To protect the PU from such interference, we consider detection decisions to be valid if the misdetection probability is lower than $\overline{MD}_{thr}$. A binary variable I_d^k is introduced to indicate whether the protection requirement is satisfied or not and is given by

$$I_d^k = \begin{cases} 1, & \text{if } 1 - F_d^k < \overline{MD}_{thr}, \\ 0, & \text{otherwise.} \end{cases} \tag{3.8}$$

If the misdetection probability of channel k exceeds $\overline{MD}_{thr}$, the detection is considered invalid, and channel k will not be allocated to data sensors. Substituting Eqs. (3.2) and (3.4) into (3.8) yields

$$I_d^k = \begin{cases} 1, & \text{if } \Pi_{m \in \mathcal{M}_k}\left(1 - Q\left(\frac{Q^{-1}(\bar{p}_f) - \sqrt{U}\gamma_{m,k}}{\sqrt{2\gamma_{m,k}+1}}\right)\right) \\ & < \overline{MD}_{thr}, \\ 0, & \text{otherwise.} \end{cases} \tag{3.9}$$

The objective of the SSS problem is to maximize the average detected available time of channels while protecting the PUs, which can be written as

$$\sum_{k=1}^{K} \alpha^k (1 - \bar{p}_f)^{\sum_{m=1}^{M}[\mathbf{J}]_{m,k}} I_d^k. \tag{3.10}$$

The SSS problem is subject to two constraints; the first constraint is related to the EH dynamics, whereas the second constraint is related to the frame structure (see Fig. 3.2). To maintain the sustainability of the spectrum sensors, the energy consumption of each sensor is upper bounded by its harvested energy. This arrangement can be mathematically written as

$$\left(\sum_{k=1}^{K} J_{m,k}\right) e_s \leq \pi_m T. \tag{3.11}$$

Moreover, the time that is spent on sensing channel k is bounded by the duration of the spectrum-sensing phase τ_s, namely,

$$\left(\sum_{m=1}^{M} J_{m,k} \right) \tau_{s'} \leq \tau_s. \tag{3.12}$$

Then, the SSS problem becomes a combinatorial problem of optimizing the sensor-to-channel assignment matrix $\mathbf{J}$ and can be written as follows:

$$(\text{SSS}) \max_{\mathbf{J}} \sum_{k=1}^{K} \alpha^k (1 - \bar{p}_f)^{\sum_{m=1}^{M}[\mathbf{J}]_{m,k}} I_d^k$$

$$\text{s.t.} \begin{cases} \left(\sum_{k=1}^{K} J_{m,k} \right) e_s \leq \pi_m T, \forall m, \\ \left(\sum_{m=1}^{M} J_{m,k} \right) \tau_{s'} \leq \tau_s, \forall m, \\ J_{m,k} = \{0, 1\} \ \forall m, k. \end{cases}$$

The actual available time of channel k, i.e., α^k, has a constant value over a given channel. As more channels are assigned to a given spectrum sensor, i.e., as $\sum_{m=1}^{M} J_{m,k}$ increases, the value of $(1 - \bar{p}_f)^{\sum_{m=1}^{M} J_{m,k}}$ decreases, and I_d^k tends to take a unit value. Therefore, there exists a trade-off between $(1 - \bar{p}_f)^{\sum_{m=1}^{M} J_{m,k}}$ and I_d^k. However, the assignment $J_{m,k}$ exists in the exponential part of $(1 - \bar{p}_f)^{\sum_{m=1}^{M} J_{m,k}}$ and affects I_d^k through $\mathcal{M}_k$. These structures make the SSS problem an integer programming problem. Intuitively, the objective function can be maximized by performing an exhaustive search over the space that is characterized by the constraints of SSS. However, this arrangement leads to a search space of size 2^{MK} which is computationally prohibitive especially for the resource-limited sensor network. In the following subsection, we apply the cross-entropy-based algorithm (C-E algorithm) [17] to address the SSS problem.

3.3.1.2 Cross-Entropy-Based Algorithm

The basic idea of the C-E algorithm is to transform a deterministic problem to the related stochastic optimization problem such that rare-event simulation techniques can be applied. More specifically, C-E algorithm first defines an associated stochastic problem for the deterministic problem, and then, solves the associated problem by an adaptive scheme. The adaptive scheme generates random solutions that converge stochastically to the optimal or near-optimal solution of the original deterministic problem.

Before introducing the C-E algorithm, we transform the constrained problem into an unconstrained problem by applying a penalty method. Let $\omega = -\sum_{k=1}^{K} \alpha^k$ be the penalty for violating any of the constraints, and then, the SSS problem transforms to

$$O = \omega \cdot I_{(\sum_{m=1}^{M} [J]_{m,k} \cdot e_s > \pi_m T)} + \omega \cdot I_{(\sum_{k=1}^{K} [J]_{m,k} \cdot \tau_{s'} > \tau_s)}$$
$$+ \sum_{k=1}^{K} \alpha^k (1 - \bar{p}_f)^{\sum_{m=1}^{M} [J]_{m,k}} I_d^k. \tag{3.13}$$

For a negative constant penalty of ω, the unconstrained objective function evaluates to a negative value for all of the infeasible solutions that violate constraints (3.11) and (3.12). The indicator function, $I_{(.)}$, takes the value of 1 for true evaluations of $(\cdot)$ and zero otherwise.

The row vectors of sensor-to-channel matrix $\mathbf{J}$ are drawn from a set, $\mathscr{C}$, of channel assignment vectors that hold a sequence of binary numbers, $\mathscr{C} = \{\mathbf{1}, \ldots, \mathbf{c}, \ldots, \mathbf{C}\}$, and the cardinality of the set is $C = |\mathscr{C}| = 2^K$. Mathematically, $[J_{m,1:K} \in \mathscr{C}$. Although the cardinality of $\mathscr{C}$ grows exponentially with K, we focus on a single-hop network with limited number of unlicensed channels; hence, the value of C is also limited. Next, the C-E algorithm allocates channel assignment vectors to the spectrum sensors rather than individual channels as in $\mathbf{J}$. Define a channel assignment vector to the spectrum sensors binary assignment matrix, $\mathbf{V}^z = \{v_{m,c}^z \mid 1 \leq m \leq M, c \in \mathscr{C}\}$, of size $M \times C$, where a value of 1 for $v_{m,c}^z$ indicates that the channel assignment vector c is allocated to spectrum sensor m. In each iteration of the C-E algorithm, random samples of this matrix are generated, and the superscript z is introduced to denote the sample number.

The $\mathbf{V}^z$ samples are generated following a probability mass function (p.m.f.) that is denoted by matrix $\mathbf{Q}^i$, which is defined as $\mathbf{Q}^i := \{q_{m,c}^i \mid 1 \leq m \leq M, c \in \mathscr{C}\}$, where $q_{m,c}^i$ denotes the probability that m is scheduled to sense the channels in vector c. The C-E algorithm operates iteratively, and in every step, the p.m.f. matrix is updated. The superscript $(\cdot)^i$ denotes the iteration number. Each iteration of the C-E algorithm consists of the following steps:

1. *Initialization*: Set the iteration counter to $i = 1$ and the maximum iteration number to $i_{\max}$. Set the initial stochastic policy of all of the spectrum sensors to be a uniform distribution on the channel assignment vector set $\mathscr{C}$, such that m chooses vector c with probability $q_{m,c}^1 = 1/C$, $\forall m, c$.
2. *Generation of Sample Solutions*: Generate Z samples of the matrix $\mathbf{V}^z$ based on the p.m.f. matrix $\mathbf{Q}^i$. Note that each spectrum sensor is randomly assigned one channel assignment vector that holds several channels, i.e., $\sum_1^C v_{m,c}^z = 1$, $\forall z \forall m$.
3. *Performance Evaluation*: Substitute the Z samples of $\mathbf{V}^z$ into Eq. (3.13) to obtain an objective function value O^z for each sample; the superscript $(\cdot)^z$ has been introduced to denote the sample number. Sort the Z values of O^z in descending order. Set ρ to be a fraction of the sorted objective values to retain, and then, take the largest $\lceil \rho Z \rceil$ values of the sorted set and ignore all of the others. Moreover, set η to be the smallest value in the sorted and retained set.
4. *p.m.f. Update*: Update the p.m.f. based on the retained objective function values. The value of $q_{m,c}^{i+1}$ is determined by

$$q_{m,c}^{i+1} = \frac{\sum_{z=1}^{Z} v_{m,c}^z I_{O^z \geq \eta}}{\lceil \rho Z \rceil}, \tag{3.14}$$

In this step, the channel vector assignment probability $q_{m,c}^i$ is updated by increasing the probability of assignments that are generating large objective function values over the various randomly generated samples.

5. *Stopping Criterion*: The algorithm stops iterating if the maximum number of iterations i_{max} is reached or the following inequality stands:

$$||\mathbf{Q}^{i+1} - \mathbf{Q}^i||_{Fr} \le \varepsilon, \qquad\qquad (3.15)$$

where $|| \cdot ||_{Fr}$ denotes the Frobenius norm.[1] Otherwise, increment the iteration counter i and go back to Step 2. Equation (3.15) represents the convergence condition of p.m.f $\mathbf{Q}^i$. It was shown in [18] that the sequence of p.m.f converges with probability 1 to a unit mass that is located at one of the samples.

Note that fine tuning the values ε impacts the convergence speed of the algorithm and the quality of the obtained solution. A large value of ε results in faster convergence but a shorter average available time of the channel.

6. *Solution Selection*: When the algorithm terminates, select the solution $\mathbf{V}^z$ that generates the largest objective value O^z. Set the values of $\mathbf{J}$ based on the assignments solution in $\mathbf{V}^z$. In other words, the assignment in $\mathbf{V}^z$ is mapped to the channel-to-sensor assignment in $\mathbf{J}$ which is a solution to the original SSS problem.

The sink schedules spectrum sensors to detect the licensed channels according to the solution obtained in Step-6 of the C-E algorithm. After the spectrum-sensing phase, spectrum sensors report their decisions on the channel availability to the sink. The sink estimates the availability of each channel based on the logic OR rule and allocates the available channels to collect data from the data sensors. In the following, we investigate the data sensor resource allocation (DSRA) problem.

3.3.2 Data Sensor Resource Allocation

Available channels detected by the spectrum sensors are allocated to the B CR transceivers that are mounted on the sink. If the number of available channels is less than B, then all of the available channels are allocated. Alternatively, the available channels are sorted with respect to their sojourn time, and the channels with the largest sojourn time values are allocated to transceivers. Let $\bar{K}$ be the number of allocated channels, and note that $\bar{K} \le B$. Because all of the channels have the same

---

[1] The Frobenius norm is defined as the square root of the sum of the absolute squares of the elements of the matrix. For example, if

$$A = \begin{bmatrix} a_{11} & a_{12} \\ a_{21} & a_{22} \end{bmatrix},$$

then

$$||A||_{Fr} = \sqrt{|a_{11}|^2 + |a_{12}|^2 + |a_{21}|^2 + |a_{22}|^2}.$$

bandwidth and average power gain, a long average sojourn time implies a large capacity.

Recall that α^k is the k-th channel's available time. However, scheduling the data sensors to transmit for the entire α^k increases the chance of collision between the data sensor and the returning PU. Let $\bar{\alpha}^k$ be the maximum access time of the k-th channel, where $\bar{\alpha}^k < \alpha^k$. It is important to design $\bar{\alpha}^k$ such that a low collision probability $p_{coll}^k(\bar{\alpha}^k)$ is maintained on the k-th channel. The probability of collision $p_{coll}^k(\bar{\alpha}^k)$ is determined in the following Theorem:

Theorem 3.1 *Given that the PU behavior on each channel is a stationary exponential ON–OFF random process, the probability of collision $p_{coll}^k(\bar{\alpha}^k)$ can be expressed by*

$$p_{coll}^k(\bar{\alpha}^k) = P_{OFF}^k \cdot (1 - e^{-\mu_k \bar{\alpha}^k}), \tag{3.16}$$

where P_{OFF}^k is the probability that PU is not present on the k-th channel at the beginning of the data transmission phase, and $(1 - e^{-\mu_k \bar{\alpha}^k})$ captures the probability that PU returns in $[0, \bar{\alpha}^k]$.

Proof Let T_{OFF}^k be the sojourn time of a OFF/Inactive period with the probability density function (p.d.f) $f_{T_{OFF}^k}(\alpha)$. Given the exponentially distributed ON/OFF period, the p.d.f of the Inactive period is equal to [14]

$$f_{T_{OFF}^k}(\alpha) = \mu_k e^{-\mu_k \alpha}.$$

The probability that the OFF/Inactive period is less than $\bar{\alpha}^k$, i.e., the PU on channel k returns in $[0, \bar{\alpha}^k]$, can be derived to be

$$Pr(T_{OFF}^k < \bar{\alpha}^k) = \int_0^{\bar{\alpha}^k} f_{T_{OFF}^k}(\alpha) \, d\alpha$$

$$= 1 - e^{-\mu_k \bar{\alpha}^k}.$$

Since channel k is available with probability P_{OFF}^k, we can have Eq. 3.16.

To maintain a target collision probability $\overline{p_{coll}^k}$, the channel access time should not exceed,

$$\bar{\alpha}^k \leq \frac{-\ln(1 - \overline{p_{coll}^k}/P_{OFF}^k)}{\mu_k}. \tag{3.17}$$

Furthermore, $\bar{\alpha}^k$ is bounded by the duration of the data transmission phase $T - \tau_s$. Thus,

$$\bar{\alpha}^k = \min\left(\frac{-\ln(1 - \overline{p_{coll}^k}/P_{OFF}^k)}{\mu_k}, T - \tau_s\right). \tag{3.18}$$

Let **T** and **P** with elements $t_{n,k}$ and $p_{n,k}$ denote the transmission time and power allocation matrices of size $N \times \bar{K}$. Let $t_{n,k}$ and $p_{n,k}$ denote the transmission time and power of the n-th data sensor over the k-th channel, respectively. The total energy consumption of the data sensors is determined by

$$\sum_{n=1}^{N} \sum_{k=1}^{\bar{K}} t_{n,k} p_{n,k}. \tag{3.19}$$

The transmission time of all of the data sensors over the k-th channel is limited by the channel access time $\bar{\alpha}^k$,

$$\sum_{n=1}^{N} t_{n,k} \leq \bar{\alpha}^k, \forall k. \tag{3.20}$$

Furthermore, the transmission time of the n-th data sensor is bounded by the duration of the data transmission phase, namely,

$$\sum_{k=1}^{\bar{K}} t_{n,k} \leq T - \tau_s, \forall n. \tag{3.21}$$

The data amount that is required from the n-th data sensor is denoted by D_n. During the data transmission phase, the n-th data sensor transmits sensed data over the k-th channel to the sink at a transmission power of $p_{n,k}$ and duration of $t_{n,k}$. The data transmission rate is given by

$$R_{n,k} = W \log_2 \left(1 + \delta_{n,k} p_{n,k} \right), \tag{3.22}$$

where $\delta_{n,k}$ represents the n-th sensor channel gain over the k-th channel at the sink. The allocated rate should be sufficiently large to support the generated data. This relationship is captured by

$$\sum_{k=1}^{\bar{K}} t_{n,k} R_{n,k} \geq D_n. \tag{3.23}$$

The transmission time $t_{n,k}$ and power $p_{n,k}$ are nonnegative. Additionally, $p_{n,k}$ is constrained by the maximum transmission power $p_{\max}$. Thus, we have

$$t_{n,k} \geq 0, \forall k, \forall n \text{ and} \tag{3.24}$$

$$0 \leq p_{n,k} \leq p_{\max}. \tag{3.25}$$

We allocate the transmission time $\mathbf{T}$ and power $\mathbf{P}$ to minimize the energy consumption of all of the data senors, which can be formulated as

$$(\text{DSRA}) \quad \min_{\mathbf{T},\mathbf{P}} \sum_{n=1}^{N} \sum_{k=1}^{\bar{K}} t_{n,k} p_{n,k}$$

$$\text{s.t.} \quad \begin{cases} \sum_{n=1}^{N} t_{n,k} \leq \bar{\alpha}^{k}, \forall k, \\ \sum_{k=1}^{\bar{K}} t_{n,k} \leq T - \tau_{s}, \forall n, \\ \sum_{k=1}^{\bar{K}} t_{n,k} W \log_{2}(1 + \delta_{n,k} p_{n,k}) \geq D_{n}, \forall n, \\ t_{n,k} \geq 0, \forall k, n, \\ 0 \leq p_{n,k} \leq p_{\max}, \forall k, n. \end{cases}$$

The amount of data to transmit is determined by the product of the transmission time $t_{n,k}$ and logarithm of the power $p_{n,k}$. These structures lead to the non-convexity of the problem DSRA with potentially multiple local optima and generally imply difficulty in determining the global optimal solution [19]. However, by showing that DSRA is biconvex in Theorem 3.2, we gain access to algorithms that efficiently solve biconvex problems [20].

Theorem 3.2 *If we fix one set of variables in* **T** *or* **P**, *then DSRA is convex with respect to the other set of variables. Thus, DSRA is biconvex.*

Proof We first determine a feasible **P**, and then, DSRA becomes a problem of determining **T** to satisfy

$$(\text{DSRA-1}) \quad \min_{\mathbf{T}} \sum_{n=1}^{N} \sum_{k=1}^{\bar{K}} t_{n,k} p_{n,k}$$

$$\text{s.t.} \quad (3.20)(3.21)(3.23)(3.24),$$

which is linear and convex due to the linear objective function and linear feasible set. DSRA-1 can be solved using the simplex method [21]. Additionally, by fixing **T**, DSRA becomes a problem of determining **P** to satisfy

$$(\text{DSRA-2}) \quad \min_{\mathbf{P}} \sum_{n=1}^{N} \sum_{k=1}^{\bar{K}} t_{n,k} p_{n,k}$$

$$\text{s.t.} \quad (3.23)(3.25).$$

DSRA-2 can be solved by the interior point method. Both DSRA-1 and DSRA-2 are convex and can be solved efficiently. Therefore, the objective function $\sum_{n=1}^{N} \sum_{k=1}^{\bar{K}} t_{n,k} p_{n,k}$ is biconvex on the feasible set which makes DSRA a biconvex problem.

3.3.2.1   Joint Time and Power Allocation (JTPA) Algorithm

Because DSRA is biconvex, the variable space is divided into two disjoint subspaces. Therefore, the problem is divided into two convex subproblems that can be solved efficiently: time allocation (DSRA-1) and power allocation (DSRA-2).

In the following, we adopt the alternate convex search in [20] to solve the DSRA problem. In every step of the proposed algorithm, one of the variables is fixed, and the other is optimized, and vice versa in the subsequent step. The proposed algorithm solves the two problems iteratively and converges to a partially optimal solution [20]. The detailed procedure of the proposed algorithm is given as follows:

Algorithm 1: Joint Time and Power Allocation (JTPA)

Data: Network parameters, stopping criterion ε and maximum number of iterations i_{max}
Result: The optimal $(\mathbf{T}^*, \mathbf{P}^*)$.

1 Choose an arbitrary starting point $(\mathbf{T}^0, \mathbf{P}^0)$ and set the iteration index as $i = 0$, and the initial solution as $z^0 = 0$;

2 **while** $z^{i+1} - z^{i-1} < \varepsilon$ *or* $i \geq i_{max}$ **do**

3   Fix $\mathbf{P}^i$ and determine the optimal $\mathbf{T}^{i+1}$ by solving DSRA-1 via the Simplex method [21];

4   Fix $\mathbf{T}^{i+1}$, determine the optimal $\mathbf{P}^{i+1}$ and objective function value z_i by solving DSRA-2 via the Interior Point method [22];

5      $i = i + 1$;

6 Return $(\mathbf{T}^{i+1}, \mathbf{P}^{i+1})$

The convergence of the proposed algorithm to the global optimum is not guaranteed since DSRA is biconvex and could have several local optima. However, because the objective function is differentiable and biconvex over a biconvex set, convergence to a stationary point that is partially optimal is guaranteed [20]. Data sensors transmit their data to the sink using the transmission time and power that is determined by the proposed JTPA algorithm.

3.4   Performance Evaluation

This section evaluates the performance of the C-E algorithm in the spectrum-sensing phase and the JTPA algorithm in the data transmission phase through simulations. The simulation results are obtained through Matlab on a computer with Intel Core(TM) i7-4510u CPU@2.00 GHz 2.6 GHz, 8 GB RAM.

We simulate an HSHSN that consists of $M = 8$ spectrum sensors and $N = 20$ data sensors. The sensors are randomly placed in a circular area with a radius of 20 m. The sink is located at the center of this circular area. The HSHSN coexists with a primary network that is deployed over an area that has a radius of 200 m. The PUs' transmission power is 1 mW, and the noise power is −80 dB. The PU's channel gain at the sensor is simulated based on $1/d^{3.5}$, where d is the distance between the PU

Table 3.1 Parameter settings

Parameter	Settings
Bandwidth of the licensed channel W [10]	6 MHz
Sampling rate of the EH spectrum sensor U [15]	6000
False alarm probability $\bar{p}_f$	0.1
Energy consumption per spectrum sensing	0.11 mJ
Time consumption per spectrum sensing $\tau_{s'}$	1 ms
Duration of the spectrum-sensing phase τ_s	5 ms
Upper bound of the collision probability $p_{coll}^k(\bar{\alpha}^k)$	0.1
Misdetection probability threshold $\overline{MD}_{thr}$	0.9
Fraction of samples retained in C-E ρ	0.6
Stopping threshold of C-E ε	10^{-3}

and spectrum sensor. The target false alarm probability for all of the spectrum sensors $\bar{p}_f$ is set to 0.1. PUs transmit QPSK-modulated signals, with each over a 6 MHz bandwidth W. The default number of licensed channels is five unless specified otherwise. Over the seven channels, seven PUs operate over one channel exclusively. Their transition rates $\lambda_k, k = 1, \ldots, 5$, are 0.6, 0.8, 1, 1.2, 1.4, respectively. Additionally, the transition rates $\mu_k, k = 1, \ldots, 5$ are 0.4, 0.6, 0.8, 1, 1.2, respectively. The network operates periodically over slots of length $T = 100$ ms [23] (see Fig. 3.2). The maximum transmission power is set to $p_{max} = 5$ mW [24]. The remaining parameters are set according to Table 3.1 unless specified otherwise. In the following two subsections, we evaluate the performances of the proposed algorithms.

3.4.1 Detected Channel Available Time

The following simulation results provide insights into the performance of the C-E algorithm over the spectrum-sensing phase. Metrics of interest include the convergence speed and quality of the obtained solution. Furthermore, we study the impact of the stopping criterion parameters ρ and ε on those metrics. The performance of the proposed algorithm is also compared to the performance of a candidate greedy algorithm.

In Fig. 3.4, we show the optimality of the C-E algorithm in a scenario that has four spectrum sensors and 3–5 licensed channels. We reduce the number of spectrum sensors in such a way that an exhaustive search can be efficiently performed. The EH rate and sensing time are set to be sufficiently large that any assignment would be feasible. The C-E algorithm's optimal solution, i.e., the detected average available time of channels (DAATC), is compared to that obtained by random assignment and exhaustive search. The random assignment randomly assigns licensed channels to the spectrum sensors, while the exhaustive search traverses all of the possible

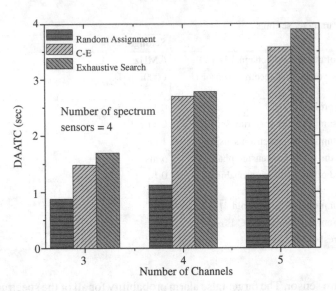

Fig. 3.4 The comparison of C-E algorithm's performance and the performance of random assignment and exhaustive search in terms of the DAATC

assignments. As shown in Fig. 3.4, the expected detected channel's available time obtained by the C-E algorithm is close to that of the exhaustive search and is able to achieve 87%–96% of it. The proposed algorithm's computed solution is 2–3 times larger than that of the random assignment.

For a network of eight spectrum sensors with five channels, the stability of the C-E algorithm is shown in Figs. 3.5 and 3.6. Figure 3.5 shows that the convergence of the C-E algorithm with respect to the EH rate ranges from 3 to 7 mW.[2] It can be seen that the value of the objective function fluctuates during the startup phase and then converges to the maximum DAATC after 20 iterations. Moreover, the value of the objective function doubles for the case in which the EH rate = 2 mW. This finding demonstrates the responsiveness of the stochastic policy updating strategy defined by Eq. (3.14). Moreover, it can be clearly seen that the DAATC increases with the EH rate.

Figure 3.6 shows the convergence results for the C-E algorithm with respect to the spectrum-sensing duration τ_s range of 2–6 ms and EH rate of 6 mW. As we can see from the figure, the value of DAATC fluctuates at the startup phase. This is because the samples of channel assignment vectors are generated according to the uniform distribution at the initialization step of the C-E algorithm. As the C-E algorithm executes, the probability to generate samples that bring higher DAATC increases. At last, the algorithm converges to a stable solution that leads to highest DAATC in 40 iterations. Furthermore, the DAATC increases with the length of the

---

[2]In [12], the real experimental data obtained from the baseline measurement system (BMS) of the Solar Radiation Research Laboratory (SRRL) shows that the EH rate ranges from 0 to 100 mW for most of the day.

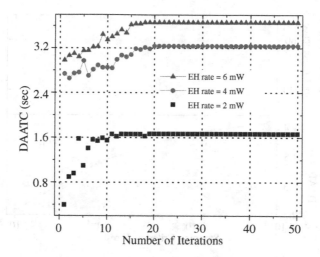

Fig. 3.5 Convergence of the C-E algorithm for three different EH rates, π_m

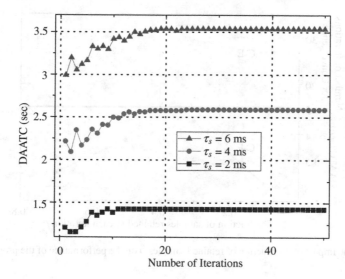

Fig. 3.6 Convergence of the C-E algorithm for three different spectrum-sensing durations, τ_s

spectrum-sensing phase τ_s, because more channels can be detected by the spectrum sensors with larger τ_s.

The C-E algorithm stops iterating if the inequality in Eq. (3.15) holds, or the maximum number of iterations is reached. Figures 3.7 and 3.8 show the impact of fine tuning the algorithm parameters, ε and ρ, on the convergence speed and quality of the obtained solution. It can be seen from Fig. 3.7 that a large number of iterations are required to satisfy the stopping criterion, and a larger DAATC can be obtained for a small ε. Furthermore, the algorithm converges in less than 80 iteration even for the

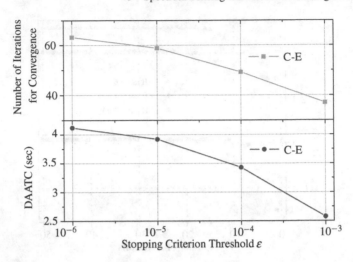

Fig. 3.7 The effect of ε on the performance of the C-E algorithm

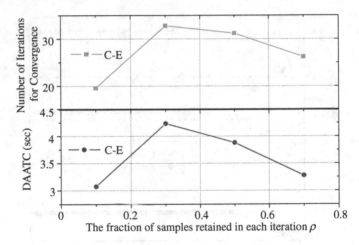

Fig. 3.8 The impact of the fraction of retained samples ρ on the performance of the proposed C-E algorithm

small value of $\varepsilon = 10^{-6}$. Figure 3.8 shows the impact of the fraction of samples that is retained (i.e., ρ) in each step on the algorithm performance. The C-E algorithm converges faster with small ρ. Moreover, the DAATC peaks at one value of ρ and then starts falling. For the parameters that considered in this study, ρ peaks at 0.3. The fraction ρ should be optimized to obtain a larger DAATC.

Figures 3.9 and 3.10 show the comparison between the performance of the C-E algorithm and that of the greedy algorithm. The greedy algorithm corresponds to the algorithm proposed in [25]; it picks the spectrum sensors sequentially and assigns them the channels that bring the largest DAATC. It can be seen from Fig. 3.9 that the

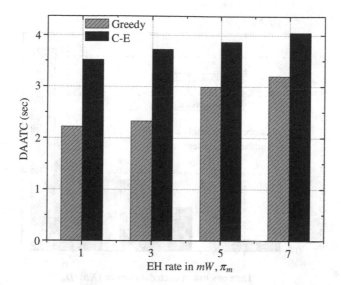

Fig. 3.9 A comparison of the C-E algorithm and the Greedy algorithm performance for a range of EH rates

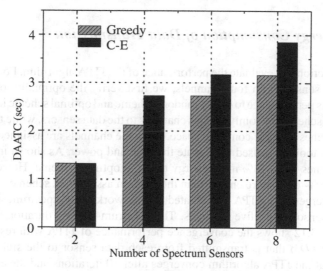

Fig. 3.10 A comparison of the C-E algorithm and the Greedy algorithm performance for a number of spectrum sensors

C-E algorithm outperforms the greedy algorithm in terms of the obtained DAATC over a range of EH rates. A similar result can be seen in Fig. 3.10, where the number of spectrum sensors varies for a fixed EH rate of 8 mW.

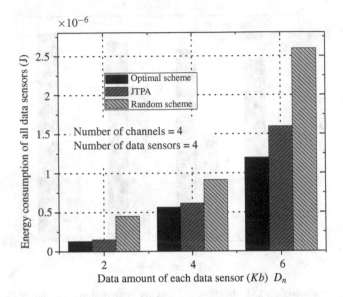

Fig. 3.11 A comparison of JTPA with the random scheme and optimal scheme

3.4.2 Energy Consumption of Data Transmission

In this subsection, we evaluate the performance of the JTPA algorithm. For a network of four data sensors with four channels, we first verify the optimality of JTPA by comparing its performance to that of random scheme and optimal scheme in Fig. 3.11. The random scheme randomly assigns channels to the data sensors, while the optimal scheme searches over the complete space. Once the channels are assigned, a Matlab optimization toolbox is used to allocate the time and power. As shown in Fig. 3.11, JTPA consumes 15%–33% more energy than the optimal scheme. However, JTPA conserves 33%–67% more energy than the random assignment scheme.

The convergence of JTPA is evaluated in a network of ten spectrum sensors and thirty data sensors with five channels. The spectrum-sensing duration τ_s is set to 5 ms. Figure 3.12 shows the convergence performance of JTPA with respect to the data amount (D_n) that is transmitted from each data sensor to the sink. It can be observed that the JTPA algorithm converges after 10 iterations and the energy consumption decreases 97% during the first 7 iterations which implies the efficiency of the proposed algorithm.

In Figs. 3.13 and 3.14, we compare the energy consumption of data transmission under the JTPA algorithm and the p_{max} scheme. In the p_{max} scheme, the data sensors transmit at the maximum available power p_{max}, and the transmission time is determined by solving the linear programming problem JTPA-1. The p_{max} scheme is comparable to the channel allocation scheme proposed in [6], in which data sensors transmit data at fixed transmission power. Figure 3.13 shows the comparison of the energy consumption performance with respect to various required amount of

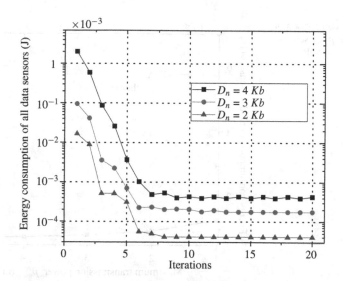

Fig. 3.12 The convergence of JTPA for a number of data amounts

Fig. 3.13 A comparison of JTPA and the p_{max} scheme for a range of data amounts

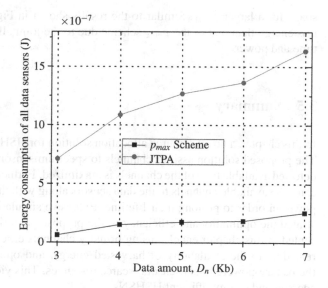

data, while p_{max} is set to 10 mW. Because the JTPA algorithm jointly allocates the transmission time and power over the available channels, JTPA consumes less energy than p_{max} scheme for different data amount. Figure 3.14 shows the comparison of the JTPA algorithm against the p_{max} scheme for various values of p_{max} and data amount $D_n = 5$ Kb $\forall n$. The energy consumption of the JTPA algorithm decreases with an increase in p_{max} because data sensors can adjust the transmission power in a larger

Fig. 3.14 A comparison of
JTPA and the p_{max} scheme
for various p_{max} values

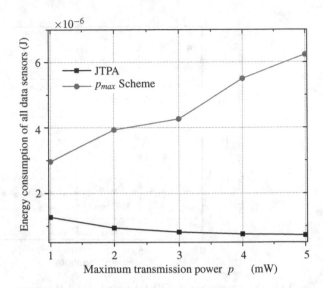

space for a larger p_{max}. Similar to the results shown in Fig. 3.13, JTPA consumes less energy than that of the p_{max} scheme due to the joint allocation of transmission time and power.

3.5 Summary

In this chapter, a novel resource allocation solution for HSHSNs has been proposed. The proposed solution assigns channels to spectrum sensors in such a way that the detected available time of the channels is maximized. Furthermore, it efficiently allocates the available channels to the data sensors along with the transmission time and power in order to prolong their lifetime. Extensive simulation results have demonstrated the optimality and efficiency of the proposed algorithms. The solution presented in this chapter enables using primary channels efficiently while adapting in real time to the availability of harvested energy, and optimizes the allocation of the battery-powered data sensors' scarce resources. This yields significantly higher spectral and energy-efficient HSHSNs.

References

1. D. Zhang, Z. Chen, J. Ren, Z. Ning, K.M. Awad, H. Zhou, X. Shen, Energy harvesting-aided spectrum sensing and data transmission in heterogeneous cognitive radio sensor network. IEEE Trans. Veh. Technol. (to be published). doi:10.1109/TVT.2016.2551721

2. P. Pratibha, K.H. Li, K.C. Teh, Dynamic cooperative sensing-access policy for energy-harvesting cognitive radio systems. IEEE Trans. Veh. Technol. (to be published). doi:10.1109/TVT.2016.2532900

3. W. Zhang, Y. Guo, H. Liu, Y. Chen, Z. Wang, J. Mitola, Distributed consensus-based weight design for cooperative spectrum sensing. IEEE Trans. Parallel Distrib. Syst. **26**(1), 54–64 (2015)

4. S.K. Nobar, K.A. Mehr, J.M. Niya, RF-powered green cognitive radio networks: architecture and performance analysis. IEEE Commun. Lett. **20**(2), 296–299 (2016)

5. N.I. Miridakis, T.A. Tsiftsis, G.C. Alexandropoulos, M. Debbah, Green cognitive relaying: Opportunistically switching between data transmission and energy harvesting. IEEE J. Sel. Areas Commun. (to be published)

6. S.-S. Byun, I. Balasingham, X. Liang, Dynamic spectrum allocation in wireless cognitive sensor networks: improving fairness and energy efficiency, in *Proceedings of IEEE VTC*, 2008, pp. 1–5

7. Z. Hu, Y. Sun, Y. Ji, A dynamic spectrum access strategy based on real-time usability in cognitive radio sensor networks, in *Proceedings of IEEE MSN*, 2011, pp. 318–322

8. J. Ayala Solares, Z. Rezki, M. Alouini, Optimal power allocation of a single transmitter-multiple receivers channel in a cognitive sensor network, in *Proceedings of IEEE ICWCUCA*, 2012, pp. 1–6

9. M. Naeem, K. Illanko, A. Karmokar, A. Anpalagan, M. Jaseemuddin, Energy-efficient cognitive radio sensor networks: Parametric and convex transformations. Sensors **13**(8), 11 032–11 050 (2013)

10. N. Zhang, H. Liang, N. Cheng, Y. Tang, J. Mark, X. Shen, Dynamic spectrum access in multi-channel cognitive radio networks. IEEE J. Sel. Areas Commun. **32**(11), 2053–2064 (2014)

11. R. Deng, J. Chen, C. Yuen, P. Cheng, Y. Sun, Energy-efficient cooperative spectrum sensing by optimal scheduling in sensor-aided cognitive radio networks. IEEE Trans. Veh. Technol. **61**(2), 716–725 (2012)

12. Y. Zhang, S. He, J. Chen, Y. Sun, X. Shen, Distributed sampling rate control for rechargeable sensor nodes with limited battery capacity. IEEE Trans. Wirel. Commun. **12**(6), 3096–3106 (2013)

13. P. Tehrani, L. Tong, Q. Zhao, Asymptotically efficient multichannel estimation for opportunistic spectrum access. IEEE Trans. Signal Process. **60**(10), 5347–5360 (2012)

14. H. Kim, K. Shin, Efficient discovery of spectrum opportunities with mac-layer sensing in cognitive radio networks. IEEE Trans. Mobile Comput. **7**(5), 533–545 (2008)

15. Y.-C. Liang, Y. Zeng, E. Peh, A.T. Hoang, Sensing-throughput tradeoff for cognitive radio networks. IEEE Trans. Wirel. Commun. **7**(4), 1326–1337 (2008)

16. S. Atapattu, C. Tellambura, H. Jiang, *Energy Detection for Spectrum Sensing in Cognitive Radio*. Springer Briefs in Computer Science (Springer, 2014)

17. R. Rubinstein, The cross-entropy method for combinatorial and continuous optimization. Methodol. Comput. Appl. Probab. **1**(2), 127–190 (1999)

18. A. Costa, O.D. Jones, D. Kroese, Convergence properties of the cross-entropy method for discrete optimization. Oper. Res. Lett. **35**(5), 573–580 (2007)

19. C. Floudas, V. Visweswaran, Comput. Chem. Eng. **14**(12), 1397–1417 (1990)

20. J. Gorski, F. Pfeuffer, K. Klamroth, Biconvex sets and optimization with biconvex functions: a survey and extensions. Math. Methods Oper. Res. **66**(3), 373–407 (2007)

21. G.B. Dantzig, M.N. Thapa, *Linear Programming 2: Theory and Extensions*, vol. 2 (2003)

22. S. Boyd, L. Vandenberghe, *Convex Optimization* (Cambridge University Press, 2004)

23. Y. Pei, Y.-C. Liang, K. Teh, K.H. Li, Energy-efficient design of sequential channel sensing in cognitive radio networks: optimal sensing strategy, power allocation, and sensing order. IEEE J. Sel. Areas Commun. **29**(8), 1648–1659 (2011)

24. T. Shu, M. Krunz, S. Vrudhula, Joint optimization of transmit power-time and bit energy efficiency in CDMA wireless sensor networks. IEEE Trans. Wirel. Commun. **5**(11), 3109–3118 (2006)

25. H. Yu, W. Tang, S. Li, Optimization of cooperative spectrum sensing in multiple-channel cognitive radio networks, in *Proceedings of IEEE GLOBECOM*, 2011, pp. 1–5

Chapter 4
Joint Energy and Spectrum Management in ESHSNs

In this chapter, we develop an aggregate network utility optimization framework for energy and spectrum management in energy and spectrum harvesting sensor networks (ESHSNs). The framework captures three stochastic processes: energy harvesting dynamics, inaccuracy of channel occupancy information, and channel fading, and does not require a priori statistics of these processes. Based on the framework, we propose an online algorithm to balance energy consumption and energy harvesting, and optimize the spectrum utilization while considering PU protection. Performance analysis shows that the proposed algorithm achieves a close-to-optimal network utility while guaranteeing network stability. Extensive simulations demonstrate the effectiveness of the proposed algorithm and the impact of network parameters on its performance.

4.1 Introduction

In the previous chapter, we propose a spectrum management scheme for a heterogeneous spectrum harvesting sensor network (HSHSN), in which the energy harvesting (EH) rate of spectrum sensors stays constant and has been known beforehand through estimation techniques. However, with energy and spectrum harvesting (ESH) capabilities, the ESHSNs are generally expected to operate for a long time duration. In this case, the availability of both harvested energy and licensed spectrum exhibits stochastic nature due to the impact of environmental factors and PU activities, which poses new challenges for resource allocation in ESHSNs. First, the EH process is stochastic and dynamic, which makes balancing energy consumption and energy replenishment challenging. Depleting a sensor's battery at a rate slower or faster than the replenishment rate leads to either energy underutilization or sensor failure, respectively [1]. Second, the spectrum utilization by sensors in ESHSNs has to adapt

D. Zhang et al., *Resource Management for Energy and Spectrum Harvesting Sensor Networks*, SpringerBriefs in Electrical and Computer Engineering, DOI 10.1007/978-3-319-53771-9_4

to the dynamic activity of PUs over the licensed spectrum [2]. For example, the spectrum occupation of cellular users is in the range of seconds or minutes [3]. When sensors transmit over the channels licensed to cellular users, the sensors may have to frequently disrupt their transmission and vacate the channels to avoid collisions with cellular users.

In the literature, the allocation of energy and spectrum in ESHSNs has been considered in [4–8] and [9–12], separately. However, the above-mentioned works fail to jointly allocate resources in ESHSNs. [13, 14] consider the coupling of resource allocation in ESHSNs. In [13], an MDP-based spectrum detection and access policy is proposed for an ESH system to maximize the throughput. Unfortunately, the complexity of the proposed policy exponentially increases with the number of nodes in the system, and thus is not practical for ESHSNs which may consist of a large number of sensors. [14] investigates the joint energy and spectrum allocation in ESHSNs for network utility optimization. However, [14] requires a perfect prior knowledge of EH process and PU activities.

To fill these research gaps, we develop an aggregate utility optimization framework to facilitate the design of an online algorithm that couples energy and spectrum access management for a single-hop ESHSN. The considered ESHSN consists of a sink and a number of sensors with ESH capabilities. The sensors harvest energy to sense data and transmit it to the sink over the unoccupied licensed spectrum. The developed framework is Lyapunov optimization-based, and captures the dynamic and stochastic nature of ESHSN resources. Based on the framework, an online algorithm is designed to achieve a close-to-optimal time-average network utility, which captures the data-sensing efficiency of the network, while ensuring protection of PUs, and a deterministic bound on sensors' data buffer and battery capacity.

This chapter is organized as follows. The detailed description of system model and problem formulation is given in Sect. 4.2. Section 4.3 presents the network utility framework and the online algorithm. Section 4.4 analyzes the stability and optimality of the proposed algorithm. Simulation results are provided in Sect. 4.4 to evaluate the performance of the proposed algorithm. Section 4.6 summarizes this chapter.

4.2   System Model and Problem Formulation

The ESHSN under consideration consists of N sensors forming the set $\mathcal{N} = \{1, 2, \ldots, N\}$ and operating over the time slots $t \in \mathcal{T} = \{0, 1, 2, \ldots\}$. The ESHSN coexists with PUs that have the privilege to access licensed channels, as shown in Fig. 4.1. The sensor collects data from an AoI and saves it in its data queue, and then transmits it to the sink over licensed channels. There are L transceivers mounted on the sink such that the sink can support L concurrent data transmission over L different frequency bands in each time slot. The availability information of the licensed spectrum is acquired from a third-party system (TPS). The TPS detects the PU activities by various existing spectrum-sensing technologies, such as energy detection [15].

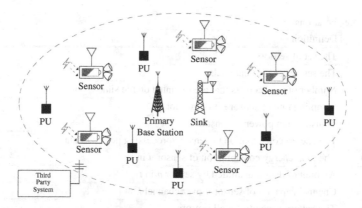

Fig. 4.1 An illustration of an ESHSN that shows the coexistence of the primary users and the sensor network

The notations used in this chapter are as follows. For a random variable X, the expected value is denoted by $\mathbb{E}[X]$, and its conditional expectation on event A is denoted by $\mathbb{E}[X|A]$. The function $[x]^+$ denotes nonnegative values, i.e., $\max(x, 0)$. The key notations are summarized in Table 4.1.

In time slot t, sensor n collects data at a sampling rate $r_n(t)$, which falls in the range:

$$0 \le r_n(t) \le r_{max}, \quad \forall n \in \mathcal{N}, \qquad (4.1)$$

where r_{max} is the maximum sampling rate. The sampling rate is associated with a utility function $U(r_n(t))$, which is increasing, continuously differentiable and strictly concave in $r_n(t)$ with a bounded first derivative $U'(r_n(t))$ and $U(0) = 0$ [16]. The concavity of the utility function is based on the observation that the marginal utility of the collected data decreases as the amount of collected data increases in sensor networks [6]. The upper bound of the first-order derivative of $U(r_n(t))$ is denoted by ζ_{max} and equals $U'(0)$.

4.2.1 Channel Allocation and Collision Control Model

The licensed spectrum consists of K orthogonal channels of equal bandwidth. The set of orthogonal channels is denoted by $\mathcal{K} = \{1, 2, \ldots, K\}$ with cardinality $K = |\mathcal{K}|$. Let $S(t) = (S_1(t), \ldots, S_K(t))$ denote the channel availability indicator with interpretation that $S_k(t) = 1$ if channel k is available, and $S_k(t) = 0$ otherwise. We assume that the PU activity on channel k evolves following an independent and identical distribution (i.i.d.) across the time slots and is uncorrelated with sensors' activities [11]. The channel unavailability rate which corresponds to the PU activity rate on channel k is given by $\beta_k = \lim_{T \to \infty} \frac{1}{T} \sum_{t=0}^{T-1} (1 - S_k(t)) \le 1$.

Table 4.1 Key notations

Notation	Definition
$\mathcal{N}$	The set of sensors
$\mathcal{K}$	The set of licensed channels
L	Number of transceivers that are mounted on the sink
$r_n(t)$	Sampling rate of sensor n in time slot t
P_T	Transmission power of sensors
P_S	The energy consumption of sensing/processing per unit data
$P_n^{Total}(t)$	The total energy consumption of sensor n in t
$x_n(t)$	Amount of data transmitted by sensor n in t
$\lambda_{n,k}(t)$	Channel capacity of sensor n over channel k in t
Ω	The battery capacity for all sensors
$e_n(t)$	Harvested energy by sensor n in t
$\eta_n(t)$	Energy supply rate of sensor n in t
$E_n(t)$	Energy queue length of sensor n in t
$Q_n(t)$	Data queue length of sensor n in t
$Z_k(t)$	Collision queue length of channel k in t
$\Theta(t)$	Conditions that impact the accuracy of channel detection in t
$Pr_k(t)$	Channel access probability of channel k in t
ρ_k	Tolerable collision rate of the PU on channel k
$J_{n,k}(t)$	Indicator of channel k assignment to sensor n in t
$S_k(t)$	Indicator of PU activity on channel k in t
$C_k(t)$	Indicator of a collision on channel k in t
ζ_U	Maximal first-order derivative of the utility function $U(r_n(t))$
η_{max}	Upper bound of energy supply rate
λ_{max}	Upper bound of channel capacity
r_{max}	Maximal sampling rate
P_{max}	Maximal energy consumption of sensors in one time slot
V	Nonnegative weight to indicate the trade-off between the network utility and queue length

The TPS provides the availability of channels to the ESHSN at the beginning of each time slot. Owing to detection errors of spectrum-sensing such as false alarms and misdetection [17], the channel availability information is assumed to be imperfect. Thus, the TPS provides channel access probability vector $\mathbf{Pr}(t) = (Pr_1(t), \ldots, Pr_k(t), \ldots Pr_K(t))$, where $Pr_k(t)$ denotes the probability that channel k is idle and hence accessible in time slot t [11]. Two factors impact the channel access probability: the actual PU activity on channel k, i.e., $S_k(t)$, and the accuracy of the spectrum-sensing techniques [15]. The performance of spectrum-sensing techniques highly depends on the receiver signal-to-noise ratio (SNR) and the detection parameters (e.g., detection threshold) [17]. These conditions in the tth time slot are collectively denoted by $\Theta(t)$. The channel access probability $Pr_k(t)$ is the conditional probability of the channel being available in time slot t, i.e., $Pr_k(t) = Pr[S_k(t) = 1|\Theta(t)]$ [11]. Because $S_k(t) = 1$ indicates that the availability of channel k, with $S_k(t) = 0$ otherwise, the closer the value of $\boldsymbol{Pr}(t)$ is to that of $\boldsymbol{S}(t)$, the more accurate the channel availability information is. An ESHSN with accurate $\boldsymbol{Pr}(t)$ is more efficient in utilizing the licensed channels by avoiding collisions.

At the beginning of each time slot, the sink allocates licensed channels to sensors based on the channel access probability. Let $\boldsymbol{J}(t)$ denote the channel allocation matrix of elements $J_{n,k}(t), \forall n \in \mathcal{N}, k \in \mathcal{K}$; $J_{n,k}(t) = 1$ if channel k is allocated to sensor n, and otherwise is 0. To avoid interference among sensors, each channel can be allocated to one sensor at most,

$$\sum_{n \in \mathcal{N}} J_{n,k}(t) \le 1, \quad \forall k \in \mathcal{K}. \tag{4.2}$$

Furthermore, each sensor can use at most one channel in each time slot, so we have

$$\sum_{k \in \mathcal{K}} J_{n,k}(t) \le 1, \quad \forall n \in \mathcal{N}. \tag{4.3}$$

Because there are L transceivers mounted on the sink, the sink can support at most L concurrent data transmissions over licensed channels in each time slot. This can be written as

$$\sum_{n \in \mathcal{N}} \sum_{k \in \mathcal{K}} J_{n,k}(t) \le L. \tag{4.4}$$

Due to the inaccuracy of channel availability and PU activities, PUs and sensors may collide over the channels. The ESHSN may access the channel that is occupied by PUs, and thus both data transmissions from PUs and sensors fail due to interference. We assume that the PU on channel k can tolerate a time-average collision rate denoted by ρ_k [11]. For example, $\rho_k = 1\%$ implies that the PU on channel k can tolerant at most 1% of data loss. Recalling that the PU on channel k is active with rate β_k, the target tolerable collision rate evaluates to $\beta_k \rho_k$. Define a collision indicator $C_k(t) \in \{0, 1\}$. The collision indicator takes a value of 1 if collision occurs and is 0 otherwise. A collision occurs when an unavailable channel is allocated to one of the sensors, such that $C_k(t) = (1 - S_k(t)) \sum_{n \in \mathcal{N}} J_{n,k}(t)$. The time-averaged rate of collision between PUs and sensors on the kth channel can be defined as

$$\bar{C}_k = \lim_{T \to \infty} \frac{1}{t} \sum_{t=0}^{T-1} C_k(t), \quad \forall k \in \mathcal{K}.$$

$\bar{C}_k$ should be less than the target-tolerable collision rate $\beta_k \rho_k$, i.e.,

$$\bar{C}_k \le \beta_k \rho_k, \quad \forall k \in \mathcal{K}. \tag{4.5}$$

To keep track of collisions between sensors and PUs, we define the virtual collision queue $Z_k(t)$ for each channel and a vector of virtual collision queues for all licensed channels, $\mathbf{Z}(t) = (Z_1(t), \ldots Z_K(t))$.

The collision queue occupancy varies following a single-server system with the collision variable $C_k(t)$ as an input process and $\rho_k 1_k(t)$ as a service process. $1_k(t)$ here is the complement of the channel availability indicator $1_k(t) = 1 - S_k(t)$. The collision queue occupancy $Z_k(t)$ evolves according to [11]:

$$Z_k(t+1) = [Z_k(t) - \rho_k 1_k(t), 0]^+ + C_k(t), \quad \forall k \in \mathcal{K}. \tag{4.6}$$

The collision queue is stable only if the time-average input rate $\lim_{t \to \infty} \frac{1}{t} \sum_{\tau=0}^{t-1} C_k(\tau) = \rho_k \beta_k$ is less than the time-average service rate $\lim_{t \to \infty} \rho_k \frac{1}{t} \sum_{\tau=0}^{t-1} (1 - S_k(\tau)) = \bar{C}_k$, i.e.,

$$\lim_{t \to \infty} \frac{1}{t} \sum_{\tau=0}^{t-1} C_k(\tau) \leq \lim_{t \to \infty} \rho_k \frac{1}{t} \sum_{\tau=0}^{t-1} (1 - S_k(\tau)),$$

which is equivalent to the constraint (4.5). Therefore, stabilizing the collision queue for each channel maintains the required PU protection.

4.2.2   Energy Supply and Consumption Model

Sensor n senses data with sampling rate $r_n(t)$ in time slot t from the AoI and saves it in the data queue. The energy consumption[1] of data sensing is assumed to be a linear function of the sampling rate $r_n(t)$ [4] and denoted by $P_S r_n(t)$. If channel k is allocated to the nth sensor, it transmits data to the sink with power $P_T, \forall n \in \mathcal{N}$. Thus, the total energy consumption P_n^{total} of the nth sensor in the tth time slot is

$$P_n^{total}(t) = P_S r_n(t) + \sum_{k \in \mathcal{K}} J_{n,k}(t) P_T, \quad \forall n \in \mathcal{N}.$$

Because the sampling rate $r_n(t)$ is bounded by r_{max} and at most one channel can be allocated to a given sensor, i.e., $\sum_{k \in \mathcal{K}} J_{n,k}(t) \leq 1$, the energy consumption is bounded by $P_n^{total}(t) \leq P_S r_{max} + P_T$. We use $P_{max} = P_S r_{max} + P_T$ to denote the upper bound of any sensor's energy consumption in a given time slot.

Sensor n is equipped with a battery of limited capacity $\Omega_n, \forall n \in \mathcal{N}$. Because the battery capacity is the same for all sensors, we omit the subscript n for simplicity. We use $E_n(t)$ to denote the energy queue length of sensor n. In time slot t, sensor

---

[1]The time is measured in unit size. Thus the implicit multiplication by 1 slot is omitted when converting between power and energy [4, 5].

n harvests energy $e_n(t)$ and consumes energy $P_n^{total}(t)$. Thus, the energy queue of sensor n evolves according to

$$E_n(t+1) = E_n(t) - P_n^{total}(t) + e_n(t). \tag{4.7}$$

In a given time slot t, the total energy consumption of sensor n must satisfy the following energy-availability constraint:

$$P_n^{total}(t) \leq E_n(t), \quad \forall n \in \mathcal{N}. \tag{4.8}$$

The energy harvesting process is characterized by the energy supply rate $\eta_n(t)$, which determines the amount of harvestable energy of sensor n in time slot t. The upper bound of $\eta_n(t)$ is denoted by $\eta_n \leq \eta_{max}$, $\forall n \in \mathcal{N}, t \in \mathcal{T}$. Furthermore, $\eta_n(t)$ randomly varies in an i.i.d fashion over slots. Notably, the exact distribution of $\eta_n(t)$ is not required, which is practically useful when knowledge of the EH process statistics is difficult to obtain. The harvested energy $e_n(t)$ is bounded by $\eta_n(t)$, i.e.,

$$0 \leq e_n(t) \leq \eta_n(t), \quad \forall n \in \mathcal{N}. \tag{4.9}$$

The total energy stored in the battery is limited by the battery capacity; thus, the following inequality must be satisfied in each time slot:

$$E_n(t) + e_n(t) \leq \Omega, \quad \forall n \in \mathcal{N}. \tag{4.10}$$

4.2.3   Data Sensing and Transmission Model

Two factors determine the amount of data that sensor n can transmit over channel k, i.e., the availability of channel k, i.e., $S_k(t)$, and the channel capacity denoted by $\lambda_{n,k}(t)$. Considering the time-varying nature of channel fading, we assume that $\lambda_{n,k}(t)$ randomly varies over time slots in an i.i.d fashion and is bounded by $\lambda_{n,k}(t) \leq \lambda_{max}$, $\forall n \in \mathcal{N}, k \in \mathcal{K}$ as in [5].

The data transmission of the sensor on channel k fails if it collides with an active PU's transmission on channel k, i.e., $S_k(t) = 0$. Denote $x_n(t)$ as the data transmission rate of sensor n in time slot t. If channel k is allocated to sensor n, the data transmission rate $x_n(t)$ is bounded by

$$x_n(t) \leq \sum_{k \in \mathcal{K}} J_{n,k}(t) S_k(t) \lambda_{n,k}(t), \quad \forall n \in \mathcal{N}. \tag{4.11}$$

Let $Q_n(t)$ denote the data queue occupancy of sensor n and $\mathbf{Q}(t) = (Q_1(t), Q_2(t), \ldots, Q_N(t))$ represent a vector of all sensors' data queue lengths. Note that $r_n(t)$ is the sampling rate, i.e., sensing rate, of sensor n in time slot t, and the dynamics of the data queue can be expressed as

$$Q_n(t+1) = Q_n(t) - \sum_{k \in \mathcal{K}} J_{n,k}(t) S_k(t) x_n(t) + r_n(t), \qquad (4.12)$$

where $J_{n,k}(t) x_n(t)$ captures the service process, whereas $r_n(t)$ models the input process. This single-server queuing system is stable if the following network-stability constraint is satisfied [18]:

$$\lim_{T \to \infty} \frac{1}{T} \sum_{t=0}^{T-1} \sum_{n \in \mathcal{N}} \mathbb{E}[Q_n(t)] < \infty. \qquad (4.13)$$

Constraint (4.13) implies that the data queues of all sensors have finite time-average occupancy.

In a given time slot, the nth sensor can only transmit the available data in its queue; hence, the following data-availability constraint must be satisfied in each time slot:

$$0 \le x_n(t) \le Q_n(t) \quad \forall n \in \mathcal{N}. \qquad (4.14)$$

4.2.4   Problem Formulation

Based on the aforementioned models, we formulate the stochastic optimization problem. The objective is to maximize the time-average aggregate network utility of ESHSNs subject to the constraints mentioned above. The time-average aggregate network utility problem can be written as

$$\bar{O} = \lim_{T \to \infty} \frac{1}{T} \sum_{t=0}^{T} -\mathbb{E}[O(t)], \qquad (4.15)$$

where $O(t) = \sum_{n \in \mathcal{N}} U(r_n(t))$ denotes the network utility in a time slot. To simplify the presentation, we use $r(t)$, $x(t)$ and $e(t)$ to denote the vectors of sampling rate $r_n(t)$, data transmission rate $x_n(t)$, and harvested energy $e_n(t)$ in time slot t, respectively. Additionally, let $\Gamma(t) \triangleq (r(t), e(t), x(t), \mathbf{J}(t))$ represent the set of these variables in time slot t.

The network utility can be maximized by optimizing $\Gamma(t)$ under the following utility maximization formulation:

$$(\textbf{UMP}) \max_{\Gamma(t)} \bar{O}$$

$$\text{s.t. Eqs. (4.1) to (4.14).}$$

In the following section, we decompose **UMP** into a series of deterministic subproblems and relax the collision constraint (4.5), network-stability constraint (4.13), and energy-availability constraint (4.8) by employing Lyapunov optimization.

4.3 Network Utility Optimization Framework

This section proposes the network utility optimization framework to facilitate the design of a low-complexity online algorithm. The proposed framework is developed on the basis of Lyapunov optimization under which the **UMP** problem is decomposed into three deterministic subproblems. This approach facilitates achieving a close-to-optimal aggregate network utility and stability, and does not require a priori knowledge of the above-mentioned stochastic processes statistics [18].

4.3.1 Problem Decomposition

The network state in time slot t is defined as $H(t) \triangleq (Z(t), Q(t), E(t), \Theta(t))$ which captures the occupancy of collision queue, data queue, and energy queue and the conditions that affect the accuracy of channel availability estimation. Define a Lyapunov function, $L(t)$, as the sum of squares of backlogs in the collision and data queues, and the spare capacity in sensors' batteries as follows:

$$L(t) = \frac{1}{2} \sum_{k \in \mathcal{K}} (Z_k(t))^2 + \frac{1}{2} \sum_{n \in \mathcal{N}} (Q_n(t))^2 + \frac{1}{2} \sum_{n \in \mathcal{N}} \left(-\hat{E}_n(t)\right)^2, \quad (4.16)$$

where $\hat{E}_n(t) = \Omega - E_n(t)$ denotes the spare capacity of the nth sensor battery. The Lyapunov function $L(t)$ can be considered a scalar measure of the congestion in $Z_k(t)$ and $Q_n(t)$, and the capacity availability in sensors' batteries. A small value of $L(t)$ indicates a low occupancy in the data and collision queues, as well as low spare capacity in energy queues $E_n(t)$, i.e., the batteries; the converse is also true. Additionally, we define the conditional Lyapunov drift as the one-slot difference of the Lyapunov function conditional on the network state, denoted by $\Delta(t) = \mathbb{E}[L(t+1) - L(t)|H(t)]$. The expectation is taken over the randomness of energy harvesting, PU activity, and channel fading, as well as the randomness in the energy management and channel allocation actions.

By minimizing $\Delta(t)$ in each time slot, the data queue $Q_n(t)$ and collision queue $Z_k(t)$ are pushed toward zero to stabilize the data queues and collision queues such that the network-stability constraint (4.13) and tolerable collision constraint (4.5) can be satisfied. Furthermore, the energy queues $E_n(t)$ are pushed toward their capacity Ω, such that sensors tend to recharge their batteries through energy harvesting. By carefully designing the value of Ω, the energy queues are guaranteed to have enough energy for data sensing and data transmission such that the energy-availability constraint (4.8) can be satisfied. The value of Ω is determined in Theorem 2 in Sect. 4.4. Thus, constraints (4.5), (4.8), and (4.13) are satisfied.

At this point, the network utility to be maximized has not yet been incorporated. Therefore, we include a weighted version of the network utility into the Lyapunov drift, and instead of minimizing $\Delta(t)$, we minimize the following drift-minus-utility $\Delta_V(t)$ function:

$$\Delta_V(t) \triangleq \mathbb{E}[\Delta(t) - VO(t)|H(t)], \tag{4.17}$$

where V is a nonnegative importance weight that represents how much we emphasize on utility maximization [18]. In other words, instead of greedily minimizing $\Delta(t)$, we minimize $\Delta_V(t)$ to jointly stabilize the queues and optimize the weighted network utility $VO(t)$. With a sufficiently large value of V, a close-to-optimal aggregate network utility can be achieved [19]. However, the data queues and energy queues become longer with a larger value of V, such that longer data queue buffers and battery capacities are required to support the ESHSN. Thus, adjusting V allows a trade-off between the reduction of queue length and optimization of the network utility.

$$D_V(t) = \sum_{n \in \mathcal{N}} \left[-\hat{E}_n(t)e_n(t) \right] + \sum_{n \in \mathcal{N}} \left[Q_n(t)r_n(t) + P_S r_n(t)\hat{E}_n(t) - VU(r_n(t)) \right]$$
$$+ \sum_{n \in \mathcal{N}} \sum_{k \in \mathcal{K}} J_{n,k}(t) \left[Z_k(t)(1 - Pr_k(t)) - (Q_n(t)x_n(t)Pr_k(t) - P_T \hat{E}_n(t)) \right]$$

$$\tag{4.20}$$

Considering that drift-minus-utility $\Delta_V(t)$ is a quadratic function of the queue lengths and variables in $\Gamma(t)$, Lemma 1 derives the upper bound of $\Delta_V(t)$. The upper bound is a linear function of the queue length and the variables in $\Gamma(t)$, which can be efficiently minimized.

Lemma 1 *Given the variables in $\Gamma(t)$, the value of $\Delta_V(t)$ is upper bounded by*

$$\Delta_V(t) \le B + \mathbb{E}[D_V(t)|H(t)], \tag{4.18}$$

where the value of constant B is independent of V and can be expressed as

$$B = \frac{N}{2} \left[(\lambda_{max})^2 + (r_{max})^2 + (P_{max})^2 + (\eta_{max})^2 \right] + \frac{1}{2}[K + \sum_{k \in \mathcal{K}} (\rho_k)^2] \tag{4.19}$$

and $D_V(t)$ is given in Eq. (4.20).

Proof By squaring both sides of Eq. (4.6), we have Eq. (4.20). Similarly, we have Eq. (4.22) from Eq. (4.12), and Eq. (4.23) from Eq. (4.7), respectively. Substituting $\mathbb{E}[C_k(t)|\Theta(t)] = \sum_{k \in \mathcal{K}} J_{n,k}(t)Pr_k(t)$ and $\mathbb{E}[1 - S_k(t)|\Theta(t)] = 1 - Pr_k(t)$ into Eq. (4.22) and rearranging the equation, we have Eq. (4.20).

$$\frac{1}{2}\left[(Z_k(t+1))^2 - (Z_k(t))^2\right]$$

$$\leq \frac{\left[(C_k(t))^2 + (\rho_k 1_k)^2 + 2Z_k(t)(C_k(t) - \rho_k 1_k)\right]}{2} \tag{4.21}$$

$$\leq \frac{1 + (\rho_k)^2}{2} + Z_k(t)(C_k(t) - \rho_k 1_k).$$

$$\frac{1}{2}\left[(Q_n(t+1))^2 - (Q_n(t))^2\right]$$

$$\leq \frac{1}{2}\left[(x_n(t))^2 + (r_n(t))^2 + 2Q_n(t)(r_n(t) - x_n(t))\right] \tag{4.22}$$

$$\leq \frac{(\lambda_{max})^2 + (r_{max})^2}{2} + Q_n(t)(r_n(t) - x_n(t))$$

$$\frac{1}{2}\left[(E_n(t+1) - \Omega)^2 - (E_n(t) - \Omega)^2\right]$$

$$\leq \frac{\left[(P_n^{total}(t))^2 + (e_n(t))^2 - 2\hat{E}_n(t)\left(e_n(t) - P_n^{total}(t)\right)\right]}{2} \tag{4.23}$$

$$\leq \frac{(P_{max})^2 + (\eta_{max})^2}{2} - 2\hat{E}_n(t)\left(e_n(t) - P_n^{total}(t)\right)$$

Rather than minimizing the drift-minus-utility $\Delta_V(t)$ function, we try to minimize its the upper bound, i.e., the right-hand side (RHS) of Eq. (4.18). Furthermore, for a given network condition $H(t)$, only $D_V(t)$ is relevant to the variables in $\Gamma(t)$. Therefore, we minimize $D_V(t)$ by solving for the optimal sampling rate $r(t)$, harvested energy $e(t)$, data transmission rate $x(t)$, and channel allocation $J(t)$ in each time slot.

4.3.1.1 Framework Structure

Exploiting the linear structure of Eq. (4.20), $D_V(t)$ can be minimized after being decomposed it into three subproblems. In particular, the three subproblems are battery management (**BM**), sampling rate control (**SRC**), and channel and data rate allocation (**CDRA**). Figure 4.2 shows the three subproblems and the data flows among them. In the following, we treat each of the subproblems separately. The subproblems **BM** and **SRC** optimize the harvested energy $e_n(t)$ and sampling rate $r_n(t)$, respectively. Both **BM** and **SRC** require local information only available at the sensor, and they can be distributively solved at each sensor. However, **CDRA** is centrally solved at the sink because it requires information on the data queue occupancy $Q(t)$, energy queue occupancy $E(t)$, and channel collision queue occupancy $Z(t)$ of all sensors. The sink gathers this information at the beginning of each time slot via the common control channel, as in [20]. In the following, each of the subproblems is solved separately.

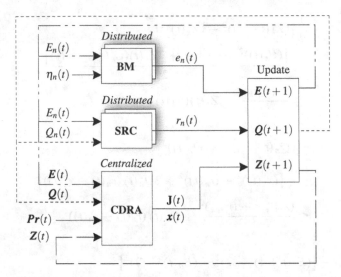

Fig. 4.2 A block diagram of the proposed framework showing the subproblems and the parameters exchanged among them

- **Battery Management**
  Considering the first term on the RHS of (4.20) and the relevant constraints (4.9) and (4.10), we have the following optimization problem to solve for $e_n(t)$

$$(\textbf{BM}) \min_{e_n(t)} -\hat{E}_n(t)e_n(t)$$

$$\text{s.t.} \begin{cases} e_n(t) \leq \eta_n(t), \\ E_n(t) + e_n(t) \leq \Omega. \end{cases}$$

If the battery is not full, i.e., $E_n(t) < \Omega$ and $\hat{E}_n(t) > 0$, the sensor should harvest as much energy as possible. Hence, if $E_n(t) < \Omega$, we have $e_n(t) = \min(\Omega - E_n(t), \eta_n(t))$; otherwise, set $e_n(t) = 0$.

- **Sampling Rate Control**
  Considering the second term on the RHS of (4.20) with constraint (4.1), we have the following optimization problem to optimize the sampling rate $r_n(t)$:

$$(\textbf{SRC}) \min_{r_n(t)} r_n(t)(Q_n(t) + P_S \hat{E}_n(t)) - VU(r_n(t))$$

$$\text{s.t. } 0 \leq r_n(t) \leq r_{max}.$$

The utility function $U(r_n(t))$ is concave; thus, the **SRC** problem is convex. Let the sampling rate $r_n^*(t)$ be the optimal solution to the **SRC** problem, based on the convex optimization theory [21], we have

$$r_n^*(t) = \left[U'^{-1} \left(\frac{Q_n(t) + P_S \hat{E}_n(t)}{V} \right) \right]_0^{r_{max}}, \quad (4.24)$$

where $[z]_a^b = \min(\max(z, a), b)$ and $U'^{-1}(\cdot)$ is the inverse of the first derivative of $U(\cdot)$.

- **Channel and Data Rate Allocation**

  Considering the third term on the RHS of (4.20) with constraints (4.2), (4.3), (4.4), (4.11), and (4.14), the problem of interest is determining the channel allocation matrix $\mathbf{J}(t)$ and data transmission rate $\mathbf{x}(t)$, which can be written as follows:

$$\textbf{(CDRA)} \quad \min_{\mathbf{J}(t), \mathbf{x}(t)} \sum_{n \in \mathcal{N}} \sum_{k \in \mathcal{K}} J_{n,k}(t) \left[Z_k(t)(1 - Pr_k(t)) \right.$$

$$\left. - (Q_n(t)x_n(t)Pr_k(t) - P_T \hat{E}_n(t)) \right]$$

$$\text{s.t.} (4.2) \ (4.3) \ (4.4) \ (4.11) \ (4.14).$$

CDRA optimizes the data transmission rate $\mathbf{x}(t)$ and channel allocation $\mathbf{J}(t)$; the former is a continuous variable and the latter is an integer variable which makes this subproblem a mixed integer problem. To facilitate the design of a tractable resource allocation solution, we transform **CDRA** into an integer problem by relaxing the constraints related to $\mathbf{x}(t)$, i.e., constraints (4.11) and (4.14). This is achieved over two steps. First, we adjust the data queue length and modify the objective function of **CDRA**; we refer to it hereafter as the modified **CDRA** (**m-CDRA**). We show that the objective function of **m-CDRA** is minimized if sensors transmit data at full capacity on their assigned channels. Second, we replace the continuous variable $x_n(t)$ by the channel capacity $\lambda_{n,k}(t)$ in the objective function of **m-CDRA**. Thus, we can relax constraints (4.11) and (4.14) and transform the **m-CDRA** problem into a channel allocation (**CA**) problem, which is a one-to-one matching problem. These steps are detailed in the following:

1. Define the adjusted length of the data queue as

$$\hat{Q}_n(t) = [Q_n(t) - \lambda_{max}]^+. \quad (4.25)$$

   After replacing $Q_n(t)$ by $\hat{Q}_n(t)$ in the objective function of the **CDRA** problem, we rewrite the objective function as

$$\sum_{n \in \mathcal{N}} \sum_{k \in \mathcal{K}} J_{n,k}(t) M_{n,k}(t), \quad (4.26)$$

   where $M_{n,k}(t) = [Z_k(t)(1 - Pr_k(t)) - (\hat{Q}_n(t)x_n(t)Pr_k(t) - P_T \hat{E}_n(t))]$. Instead of solving the original **CDRA**, we solve the modified **m-CDRA** with Eq. (4.26) as the objective function to find a suboptimal solution for the original **CDRA**. Suppose that $\mathbf{J}^*(t)$ and $\mathbf{x}^*(t)$ are the optimal solutions for the **m-CDRA**; in the following lemmas, we show that a channel k is assigned to sensor n if and only if it has a sufficient amount of data to transmit and it transmits it at full channel capacity.

Lemma 2 *For a channel k to be assigned to sensor n, i.e., $\sum_{k \in \mathcal{K}} J_{n,k}^*(t) = 1$, the following inequality must be satisfied:*

$$Q_n(t) > \lambda_{max}. \tag{4.27}$$

Proof First, we prove that if there is any channel assigned to sensor n, its adjusted queue length $\hat{Q}_n(t) > 0$. Suppose that channel k is assigned to sensor n, i.e., $J_{n,k}^*(t) = 1$, it is obvious that $M_{n,k}(t) = Z_k(t)(1 - Pr_k(t)) + P_T \hat{E}_n(t) - \hat{Q}_n(t) x_n(t) Pr_k(t) < 0$. Since $Z_k(t)(1 - Pr_k(t)) \geq 0$, $P_T \hat{E}_n(t) \geq 0$, and $x_n(t) Pr_k(t) \geq 0$, the adjusted data queue length becomes larger than zero, i.e., $\hat{Q}_n(t) > 0$.

Then, we prove that if $\hat{Q}_n(t) > 0$, then the queue length $Q_n(t) > \lambda_{max}$. Recalling that $\hat{Q}_n(t) = \max(Q_n(t) - \lambda_{max}, 0)$, $\hat{Q}_n(t) > 0$ implies that $Q_n(t) - \lambda_{max} > 0$, i.e., the data queue length $Q_n(t)$ is larger than the maximum channel capacity λ_{max}.

Lemma 3 *If any channel is assigned to the nth sensor in the tth time under the modified m-CDRA, i.e., $\sum_{k \in \mathcal{K}} J_{n,k}^*(t) = 1$, then we have*

$$x_n^*(t) = \sum_{k \in \mathcal{K}} J_{n,k}^*(t) \lambda_{n,k}(t), \tag{4.28}$$

otherwise, $x_n^*(t) = 0$.

Proof We first consider the condition that sensor n is not assigned with any channel, thus $J_{n,k} = 0$, $\forall k \in \mathcal{K}$. According to constraint (4.11) and $x_n(t) \geq 0$, we have $x_n^*(t) = 0$.

Next, we prove that $x_n^*(t) = \sum_{k \in \mathcal{K}} J_{n,k}^*(t) \lambda_{n,k}(t)$ is the optimal solution for the condition that $\sum_{k \in \mathcal{K}} J_{n,k}^*(t) = 1$. We use k_n to denote the channel assignment to sensor n, i.e., $J_{n,k_n}^* = 1$. Since M_{n,k_n} is inversely correlated with the value of $x_n(t)$, the value of $x_n(t)$ should be as large as possible to minimize M_{n,k_n}. The value of $x_n(t)$ is bounded by constraints (4.11) and (4.14), i.e., the channel capacity and data queue length $Q_n(t)$. According to Lemma 2, we can see that the data queue length $Q_n(t)$ must exceed the maximum channel capacity ($Q_n(t) \geq \lambda_{max}$) if $\sum_{k \in \mathcal{K}} J_{n,k}^*(t) = 1$. Therefore, $x_n(t)$ is only bounded by channel capacity constraint in (4.11). Then we have the optimal $x_n(t)$ to be $x_n^*(t) = \lambda_{n,k_n}(t)$.

2. Lemma 3 shows that the sensor must fully utilize the assigned channel to optimally solve **m-CDRA**. Therefore, we can replace the transmission rate $x_n(t)$ by channel capacity $\lambda_{n,k}(t)$ in Eq. (4.26) and, thus, relax the channel capacity constraint (4.11) and data-availability constraint (4.14). The modified **m-CDRA** is transformed into a **CA** problem as follows:

$$\textbf{(CA)} \min_{\mathbf{J}(t)} \sum_{n,k} J_{n,k}(t) \left[Z_k(t)(1 - Pr_k(t)) \right.$$

$$\left. - \left(\hat{Q}_n(t)\lambda_{n,k}(t)Pr_k(t) - P_T\hat{E}_n(t) \right) \right]$$

s.t. (4.2) (4.3) (4.4).

CA can be mapped to a one-to-one matching problem. Furthermore, due to the limited number of transceivers on the sink, i.e., $L \leq K$, a maximum of L channels can be allocated to sensors in a given time slot. Meanwhile, if $L < K$, i.e., not all channels can be allocated to the sensors, **CA** is an unbalanced matching problem, which can be solved by the adaptive Hungarian algorithm proposed in [22]. The complexity of the algorithm increases linearly with the number of sensors.

4.3.2 Utility-Optimal Resource Management Algorithm

In this subsection, we present the UoRMA algorithm in Algorithm 1. The UoRMA algorithm achieves the optimal harvested energy $e^*(t)$, sampling rate $r^*(t)$, data transmission rate $x^*(t)$, and channel allocation $\mathbf{J}^*(t)$ by solving **BM**, **SRC**, and **CDRA**, respectively. Moreover, the occupancies of data queues $Q(t)$, energy queues $E(t)$, and collision queues $Z(t)$ are updated according to their respective queue dynamics.

Both the **BM** and **SRC** problems have closed-form solutions, and can be distributively solved at each sensor. Thus, their complexity is negligible. The complexity of Algorithm 1 is dominated by solving the **CA** problem in step 8 with time complexity of $O(NKL + L^2 \log(\min(N, K)))$ [22]. Therefore, the complexity of UoRAM increases linearly with the number of sensors N. Notably, the complexity of algorithms designed based on Markov decision process (MDP) increases exponentially with N [23]. Comparing to the MDP-based algorithms, UoRMA is more computationally efficient in addition to being scalable for densely deployed sensor networks.

4.4 System Performance Analysis

In this section, we analyze the stability and performance of the proposed UoRMA algorithm. Theorem 1 proves the stability of ESHSNs operating under the UoRMA algorithm by deriving upper bounds on the length of the data queues and collision queues. Then, we derive the required battery capacity to support the operation of the ESHSN in Theorem 2. Theorem 3 evaluates the gap between the network's aggregate utility obtained by UoRMA and the optimal solution to demonstrate the optimality of UoRMA.

4.4.1 Upper Bounds on Queues

We derive the upper bounds on the occupancies of queues and collision queues in Theorem 1. The existence of the bounds guarantees satisfying the data and collision queue stability constraints (4.13) and (4.5).

Theorem 1 *For a nonnegative parameter V, $P_k(t) \leq 1 - \varepsilon$, $\forall k, t$, and an initialization of the collision queue and data queue satisfying $0 \leq Z_k(0) \leq Z_{max}$, $\forall k \in \mathcal{K}$ and $0 \leq Q_n(0) \leq Q_{max}$, $\forall n \in \mathcal{N}$, where the upper bounds are given by*

$$Q_{max} = \zeta_U V + r_{max},$$
$$Z_{max} = \frac{Q_{max} \lambda_{max} (1 - \varepsilon)}{\varepsilon} + 1,$$

we have

$$0 \leq Q_n(t) \leq Q_{max}, \ \forall n \in \mathcal{N}, \tag{4.29}$$
$$0 \leq Z_k(t) \leq Z_{max}, \ \forall k \in \mathcal{K}. \tag{4.30}$$

Algorithm 2: Proposed UoRMA algoritm

Data: $Z(t)$, $Q(t)$, $E(t)$, $Pr(t)$, $\eta_n(t)$, $\forall n \in \mathcal{N}$, $\lambda_{n,k}(t)$, $\forall n \in \mathcal{N}$, $\forall k \in \mathcal{K}$.
Result: $r^*(t)$, $e^*(t)$, $x^*(t)$, $\mathbf{J}^*(t)$, $Z(t+1)$, $Q(t+1)$, $E(t+1)$.

```
   /* Battery Management                                              */
1  foreach n ∈ 𝒩 do
2   │  if Eₙ(t) < Ω then
3   │  │   eₙ*(t) = min(Ω − Eₙ(t), ηₙ(t));
4   │  else
5   │  └   eₙ*(t) = 0;

   /* Sampling Rate Control                                           */
6  foreach n ∈ 𝒩 do
7   └  Compute rₙ*(t) based on Eq. (4.24);

   /* Channel and Data Rate Allocation                                */
8  Solve CA problem and set J*(t);
9  foreach n ∈ 𝒩 do
10  │  if ∑ₖ∈𝒦 Jₙ,ₖ*(t) == 1 then
11  │  │   xₙ*(t) = ∑ₖ∈𝒦 Jₙ,ₖ*(t)λₙ,ₖ(t);
12  │  else
13  │  └   xₙ*(t) = 0;

   /* Update the queue lengths                                        */
14 foreach n ∈ 𝒩 do
15  │  Compute Qₙ(t + 1) based on Eq. (4.12);
16  └  Compute Eₙ(t + 1) based on Eq. (4.7);

17 foreach k ∈ 𝒦 do
18  └  Compute Zₖ(t + 1) based on Eq. (4.6);
```

Proof When $t = 0$, Eq. (4.29) holds. In the following, we prove Eq. (4.29) by inductions. We first assume that Eq. (4.29) holds in time slot t, and then prove that it holds in $t + 1$.

1. If sensor n does not sense any data, then we have $Q_n(t + 1) \leq Q_n(t) \leq \zeta_U V + r_{max}$;
2. If sensor n collects data with sampling rate $r_n^*(t)$, given in Eq. (4.24), then we have $V U'(r_n^*(t)) = Q_n(t) - P_S(E_n(t) - \Omega)$ and $Q_n(t) \leq V U'(r_n^*(t))$. Since $U'(r_n^*(t)) \leq \zeta_U$, $\forall r_n(t)$ where ζ_U denotes the upper bound of the first-order derivative of $U(r_n(t))$, $\forall r_n(t)$, we have $Q_n(t) \leq V \zeta_U$. Furthermore, since $r_n^*(t) \leq r_{max}$, we have $Q_n(t + 1) \leq Q_n(t) + r_{max} \leq V \zeta_U + r_{max}$.

Summarily, we have $Q_n(t + 1) \leq V \zeta_U + r_{max}$. This completes the proof of Eq. (4.29).

Then we prove Eq. (4.30) by inductions. At $t = 0$, the collision queue is initialized as an empty queue. We prove that if Eq. (4.30) holds in time slot t, it will hold in $t + 1$.

1. If $P_k(t) = 1$, then no collision can happen, such that $Z_k(t + 1) \leq Z_k(t) \leq Z_{max}$.
2. If $P_k(t) \leq 1 - \varepsilon$, and $Z_k(t) \leq Z_{max} - 1$, then we have $Z_k(t + 1) \leq Z_k(t) + 1 \leq Z_{max}$.
3. If $P_k(t) \leq 1 - \varepsilon$, and $Z_k(t) > Z_{max} - 1$, then we have $Z_k(t)(1 - Pr_k(t)) - P_T(E_n(t) - \Omega) - Q_n(t)x_n(t)Pr_k(t)) \geq 0$, so channel k cannot be allocated to any sensor in problem **CA**. This would yield $C_k(t) = 0$. Therefore, we have $Z_k(t + 1) \leq Z_k(t) \leq Z_{max}$.

Summarily, we have $Z_k(t + 1) \leq Z_{max}$. This completes the proof of Eq. (4.30).

As we can see from Eqs. (4.29) and (4.30), both the upper bounds of data queues and collision queues increase linearly with the weight V. Since a larger V can bring higher network utility, the linear increase of upper bound on data queues indicates that a longer data buffer is required at each sensor to achieve better network performance. Furthermore, the increase of upper bound on collision queues also indicates that the PUs may experience more collisions from the ESHSN. However, the collision constraint can still be satisfied due to the existence of the upper bound on collision queues.

4.4.2 Required Battery Capacity

In Theorem 2, we determine the required battery capacity Ω in such a way that the sensor does not sense or transmit any data if the available energy is less than the maximum energy consumption of each sensor, i.e., $E_n(t) \leq P_{max}$. Therefore, the energy-availability constraint (4.8) becomes implicit.

Theorem 2 *Under the proposed framework and with a battery capacity Ω given by*

$$\Omega = \max \left(\frac{V\zeta_U}{P_S} + P_{max}, \frac{Q_{max}\lambda_{max}}{P_T} + P_{max} \right), \ \forall n \in \mathcal{N}, \qquad (4.31)$$

sensor n does not sense data or is not allocated a channel, i.e., $r_n(t) = 0$ and $\sum_{k \in \mathcal{K}} J_{n,k}(t) = 0$, if the energy queue length in a given time slot is less than the upper bound of the sensor's energy consumption, i.e., $E_n(t) < P_{max}$.

Proof We first derive an expression for Ω in such a way that sensor n does not sense data, i.e., $r_n(t) = 0$ if $E_n(t) < P_{max}$. The sampling rate $r_n(t)$ is determined by Eq. (4.24). The utility function $U(r_n(t))$ is concave; therefore, $U'^{-1}(r_n(t))$ and $r_n(t)$ are inversely proportional. Based on Eq. (4.24), sensor n does not sense any data, i.e., the sampling rate is $r_n(t) = 0$, if

$$\frac{Q_n(t) + P_S \hat{E}_n(t)}{V} \geq \zeta_U \geq U'(0). \qquad (4.32)$$

Recall that $\hat{E}_n(t) = \Omega - E_n(t)$ and rearrange Eq. (4.32) to $\Omega \geq \frac{V\zeta_U}{P_S} + E_n(t)$. To satisfy that the sensor cannot sense any data when $E_n(t) < P_{max}$, Ω can be set as follows $\Omega \geq \frac{V\zeta_U}{P_S} + P_{max}$.

Then we derive the value of Ω in such a way that no channel can be allocated to sensor n, i.e., $\sum_{k \in \mathcal{K}} J_{n,k}(t) = 0$, if $E_n(t) < P_{max}$. As we can see from the objective function of **CA**, no channel can be allocated to n if

$$Z_k(t)(1 - Pr_k(t)) + P_T \hat{E}_n(t) - \hat{Q}_n(t)\lambda_{n,k}(t)Pr_k(t) \geq 0. \qquad (4.33)$$

Rearrange Eq. (4.33) to

$$\Omega \geq \frac{\hat{Q}_n(t)\lambda_{n,k}(t)Pr_k(t) - Z_k(t)(1 - Pr_k(t))}{P_T} + E_n(t). \qquad (4.34)$$

Since $Pr_k \leq 1$, $\hat{Q}_n(t) \leq Q_{max}$, $Z_k(t) \geq 0$ and $\lambda_{n,k}(t) \leq \lambda_{max}$, we can change the RHS of Eq. (4.34) to $Q_{max}\lambda_{max}/P_T + E_n(t)$. To guarantee that no channel can be allocated to sensor n if $E_n(t) < P_{max}$, Ω can be set to $\Omega \geq \frac{Q_{max}\lambda_{max}}{P_T} + P_{max}$. Theorem 2 is thus proved.

The required battery capacity in (4.31) is determined by both the transmission power P_T and the sensing/processing power P_S because both data arrival and departure consume energy in ESHSNs.

4.4.3 Optimality of the Proposed Algorithm

In Theorem 3, the optimality of the UoRMA algorithm is analyzed.

Theorem 3 *Suppose that the optimal network utility that can be achieved by an exact and optimal algorithm is O^* and that the network utility $\bar{O}$ achieved by the UoRMA algorithm satisfies*

$$\bar{O} \geq O^* - \frac{\tilde{B}}{V}, \tag{4.35}$$

where $\tilde{B} = B + NK(\lambda_{max})^2$.

Proof We prove the theorem by comparing the Lyapunov drift with a stationary and randomized algorithm denoted by Π. We introduce superscript Π to variables $r^{\Pi}(t)$, $e^{\Pi}(t)$, $\mathbf{J}^{\Pi}(t)$, and $P_n^{total,\Pi}(t)$ to indicate that these variables are generated under algorithm Π. Since all of the PU activities, channel condition, and EH process change in i.i.d manners across the time slots, according to Theorem 4.5 in [18], algorithm Π can yield

$$\mathbb{E}\left[\sum_{n \in \mathcal{N}} U(r_n^{\Pi}(t))\right] \leq O^* + \delta, \tag{4.36}$$

$$\left|\mathbb{E}\left[\sum_{k \in \mathcal{K}} \left(C_k^{\Pi}(t) - \rho_k(1 - S_k(t))\right)\right]\right| \leq \varrho_1\delta, \tag{4.37}$$

$$\left|\mathbb{E}\left[\sum_{n \in \mathcal{N}} \left(r_n(t) - \sum_{k \in \mathcal{K}} J_{n,k}^{\Pi}(t)x_n(t)S_k(t)\right)\right]\right| \leq \varrho_2\delta, \tag{4.38}$$

$$\left|\mathbb{E}\left[\sum_{n \in \mathcal{N}} \left(e_n^{\Pi}(t) - P_n^{total,\Pi}(t)\right)\right]\right| \leq \varrho_3\delta, \tag{4.39}$$

where $\delta > 0$ can be arbitrarily small, and ϱ_1, ϱ_2 and ϱ_3 are constant scalars.

In each time slot, the UoRAM algorithm minimizes the right-hand side of the Lyapunov drift in Eq. (4.40)

$$\tilde{D}_V(t) = \sum_{k \in \mathcal{K}} Z_k(t)\left(C_k(t) - \rho_k(1 - S_k(t))\right) - \sum_{n \in \mathcal{N}} \hat{E}_n(t)\left(e_n(t) - P_n^{total,\Pi}(t)\right)$$
$$- \sum_{n \in \mathcal{N}} (VU(r_n(t)) - Q_n(t)r_n(t)) - \sum_{n \in \mathcal{N}}\sum_{k \in \mathcal{K}} J_{n,k}(t)x_n(t)S_k(t)\hat{Q}_n(t). \tag{4.40}$$

The proof of Eq. (4.40) can be obtained by Theorem 2 in [5]. Note that $\Delta(t) - V\mathbb{E}[\sum_{n \in \mathcal{N}} U(r_n(t))] \leq \tilde{B} + \mathbb{E}[\tilde{D}_V(t)|H(t)]$, where $\tilde{B} = B + NK(\lambda_{max})^2$ is a constant w.r.t. the variables, we can have the following inequality:

$$\Delta(t) - V\mathbb{E}\left[\sum_{n\in\mathcal{N}} U(r_n(t))\right]$$

$$\leq \tilde{B} + \mathbb{E}\left[\tilde{D}_V^{UoRMA}(t)|\boldsymbol{H}(t)\right] \tag{4.41}$$

$$\leq \tilde{B} + \mathbb{E}[\tilde{D}_V^{\Pi}(t)]$$

$$\leq \tilde{B} + (\varrho_1 + \varrho_2 + \varrho_3)\delta + O^* + \delta,$$

where $\tilde{D}_V^{UoRMA}(t)$ and $\tilde{D}_V^{\Pi}(t)$ denote the value of $\tilde{D}_V(t)$ obtained under UoRMA algorithm and algorithm Π, respectively. By setting δ to zero, we can have

$$\Delta(t) - V\mathbb{E}\left[\sum_{n\in\mathcal{N}} U(r_n(t))\right] \leq O^* + \tilde{B}. \tag{4.42}$$

Taking the expectation on both sides of (4.42), summing up the equations for $t \in \mathcal{T}$, dividing by T, and letting $T \to \infty$, we have $\bar{O} \geq O^* - \tilde{B}/V$. Theorem 3 is thus proved.

If we do not transform **CDRA** to **CA**, then the gap between the solution obtained by the proposed algorithm and the optimal solution can be determined by B/V [18], where B is the constant defined in Lemma 1. Thus, the performance loss caused by the transformation is shown in $\tilde{B}$, which is larger than B. However, by Theorem 3, we see that the UoRMA algorithm can achieve an aggregate network utility within $\mathscr{O}(1/V)$ of the optimal utility without a priori knowledge of the statistics of the stochastic processes such as channel fading, PU activities, and energy harvesting.

4.5 Performance Evaluation

This section provides simulation results to evaluate the performance of the UoRMA algorithm in ESHSNs. The simulated ESHSN is randomly deployed in a circular area with a radius of 30 m and consists of $N = 12$ sensors. The sink has $L = 3$ transceivers, and is located at the center of this circular area. Similar to [4] and [5], we define a concave utility function $U(r_n(t)) = \log(1+r_n(t))$, $\forall n \in \mathcal{N}$ and, $\zeta_U = 1$. The ESHSN operates on $K = 5$ licensed channels. The energy consumption rate of data sensing $P_S = 0.1$, and the maximum sampling rate $r_{max} = 2$. The maximum energy supply rate is set to $\eta_{max} = 2$, while the energy supply rate $\eta_n(t)$, $\forall n \in \mathcal{N}$ is uniformly distributed in $[0, \eta_{max}]$.

The PU on channel k, $\forall k \in \mathcal{K}$, is inactive with probability 0.4 in each time slot. Given that PU on channel k is inactive in time slot t, the channel access probability $Pr_k(t) = 0.85$; otherwise, $Pr_k(t) = 0.15$, i.e., the misdetection and false alarm probabilities are 0.15 [3]. The tolerable collision rate ρ_k, $\forall k \in \mathcal{K}$ is set to 0.05 [11].

The channel capacity $\lambda_{n,k}(t) = \log(1 + \frac{P_T h_{n,k}(t)}{d_n^4 N_0})$, where d_n denotes the distance between sensor n and the sink, noise power $N_0 = 10^{-5}$, and the transmission power $P_T = 1$. Furthermore, the channel fading coefficients $h_{n,k}(t)$ are uniformly distributed between $(0.5, 1.5)$ and i.i.d across time slots. The upper bound of the channel capacity is $\lambda_{max} = 2$ [4]. The energy queue is initialized as in Eq. (4.31) in time slot $t = 0$, whereas the data queue and collision queue are empty at $t = 0$. The length of the simulation is set to $|\mathcal{T}| = 1 \times 10^4$.

4.5.1 Network Utility and Queue Dynamics

In Fig. 4.3, we evaluate the network utility versus the value of V ranging from 10^4 to 8×10^4. The figure shows that the network utility increases with increase V. However, the rate at which the network utility increases decreases with a larger V. When the value of V reaches 6×10^4, the network utility converges to 19.02. This is expected because the network utility is a concave function of V, as shown in Eq. (4.42). We take a large value of V to illustrate the optimal network utility ($V = 10^6$ in our setting). We compare the network utility obtained by V ranging from 10^4 to 8×10^4 to the network utility obtained by $V = 10^7$. As shown in the figure, the network utility of $V = 10^6$ is equal to 19.02. Therefore, the network utility of $V = 6 \times 10^4$ achieves the optimal value.

Figure 4.4 shows the data queue occupancy over 10,000 slots for different values of V. The time-average lengths of data queues increase with the value of V.

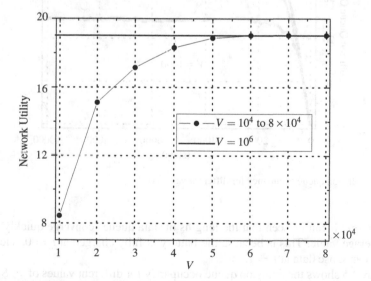

Fig. 4.3 Network utility versus V

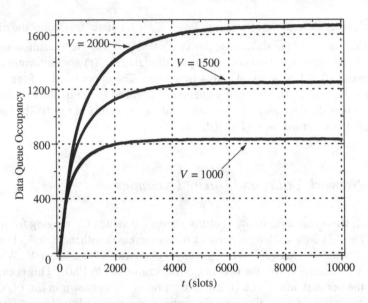

Fig. 4.4 Data queue occupancy for different values of V

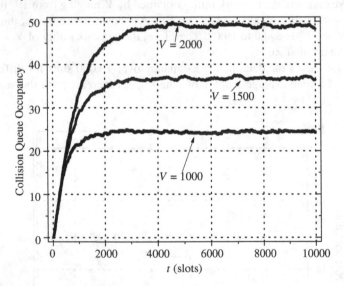

Fig. 4.5 Collision queue occupancy for different values of V

Furthermore, it can be seen that the lengths of data queues converge quickly to the time-average value. This is because the battery is fully charged at $t = 0$, such that sensors can sense data at $t = 1$.

Figure 4.5 shows the collision queue occupancy for different values of V. Similar to the data queue dynamics shown in Fig. 4.4, the time-average lengths of the collision queues increase with larger values of V, and the lengths of the collision queues fluctuate around a time-average value after the convergence. When the collision queue is small, the UoRMA algorithm tends to allocate the channel to sensors for data transmission. If the allocated channel is actually occupied by PUs, the collision queue increases back to the time-average value. Therefore, the collision queue length affects the dynamics of the queue's fluctuation. In addition, sensors' data queues and energy queues lengths also affect the dynamics of the fluctuation, because the UoRMA algorithm tends to allocate channels to the sensors with long data queues and small spare capacity in the energy queues.

4.5.2 Impact of Parameter Variation

In the following, we evaluate the impacts of various system parameters on the network utility. Assuming all channels have the same PU inactivity probability ranging from 0.4 to 0.8, we first verify the network utility in Fig. 4.6. The figure shows that the network utility increases with increase in the PU inactivity probability. At the same time, the rate of increase in the network utility decreases with higher PU inactivity probability. This is because the network utility is also limited by the energy supply rate.

Figure 4.7 shows the network utility versus maximum available energy supply η_{max} ranging from 1 to 3. The network utility monotonically increases with η_{max} because more energy can be used to sense and transmit data. However, similar to Fig. 4.6, the growth rate of the network utility decays with higher η_{max}. This indicates

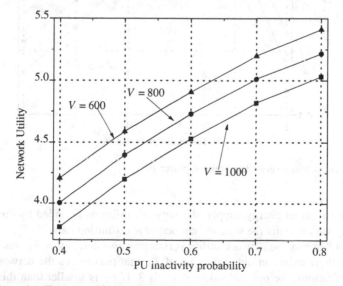

Fig. 4.6 Network utility versus PU inactivity probability

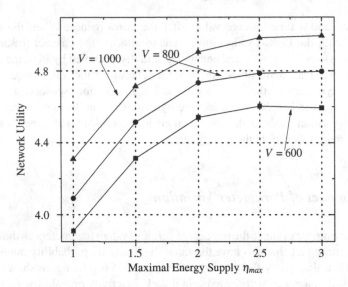

Fig. 4.7 Network utility versus maximal energy supply η_{max}

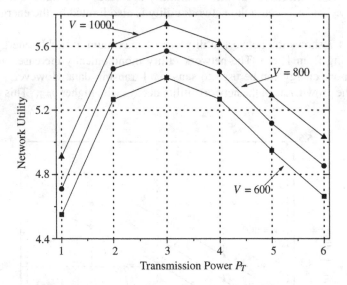

Fig. 4.8 Network utility versus transmission power P_T

that, given sufficient energy supply, the network utility is bounded by the channel availability which limits the sensors' chance of transmitting data.

Figure 4.8 shows the network utility versus transmission power P_T. As shown in the figure, there exists an optimal value of P_T that maximizes the network utility. In our simulations, the optimal value of P_T is 3. If P_T is smaller than this optimal value, the available channels are underutilized which leads to lower network utility.

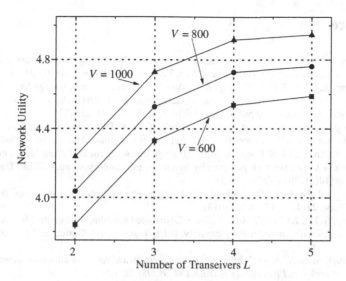

Fig. 4.9 Network utility versus number of transceivers L

However, if P_T is larger than this optimal value, sensors need more time to harvest energy for data transmission, which also reduces the network utility.

Figure 4.9 shows the network utility versus the number of transceivers that are mounted on the sink, L. Since the sink can support more concurrent data transmission with more transceivers, the network utility increases with L when $L \leq K$, i.e., number of transceivers is not larger than the number of licensed channels.

4.6 Summary

In this chapter, we have developed an aggregate network utility optimization framework to facilitate the design of an online and low-complexity algorithm for managing and allocating the resources of ESHSNs. The proposed framework captures and optimizes stochastic energy harvesting and consumption processes, as well as stochastic spectrum utilization and access processes. We employ Lyapunov optimization to decompose the problem into three subproblems that are easier to solve, battery management, sampling rate control, and data rate and channel allocation. The solutions proposed to solve the three problems constitute the proposed UoRMA algorithm. The optimality gap and bounds on data and energy queues are derived. The proposed algorithm achieves a close-to-optimal aggregate network utility while ensuring bounded energy and date queues. Simulations verify the optimality and stability of ESHSNs when operating under UoRMA algorithm. The outcomes of this chapter can be used to guide the design of practical ESHSNs by guaranteeing PU protection and sensors sustainability.

References

1. J. Zheng, Y. Cai, N. Lu, Y. Xu, X. Shen, Stochastic game-theoretic spectrum access in distributed and dynamic environment. IEEE Trans. Veh. Technol. **64**(10), 4807–4820 (2015)
2. D. Zhang, Z. Chen, J. Ren, Z. Ning, K.M. Awad, H. Zhou, X. Shen, Energy harvesting-aided spectrum sensing and data transmission in heterogeneous cognitive radio sensor network. IEEE Trans. Veh. Technol. (to be published). doi:10.1109/TVT.2016.2551721
3. N. Zhang, H. Liang, N. Cheng, Y. Tang, J. Mark, X. Shen, Dynamic spectrum access in multi-channel cognitive radio networks. IEEE J. Sel. Areas Commun. **32**(11), 2053–2064 (2014)
4. W. Xu, Y. Zhang, Q. Shi, and X. Wang, Energy management and cross layer optimization for wireless sensor network powered by heterogeneous energy sources, IEEE Trans. Wirel. Commun. **14**(5), 2814–2826 (2015)
5. L. Huang, M. Neely, Utility optimal scheduling in energy-harvesting networks. IEEE/ACM Trans. Netw. **21**(4), 1117–1130 (2013)
6. Y. Zhang, S. He, J. Chen, Y. Sun, X. Shen, Distributed sampling rate control for rechargeable sensor nodes with limited battery capacity. IEEE Trans. Wirel. Commun. **12**(6), 3096–3106 (2013)
7. R.-S. Liu, P. Sinha, C. Koksal, Joint energy management and resource allocation in rechargeable sensor networks, in *Proceedings of IEEE INFOCOM* (2010)
8. Y. Zhang, S. He, J. Chen, Data gathering optimization by dynamic sensing and routing in rechargeable sensor networks. IEEE/ACM Trans. Netw. **24**(3), 1632–1646 (2016)
9. H. Li, X. Xing, J. Zhu, X. Cheng, K. Li, R. Bie, T. Jing, Utility-based cooperative spectrum sensing scheduling in cognitive radio networks. IEEE Trans. Veh. Technol. (to be published). doi:10.1109/TVT.2016.2532886
10. M. Ozger, E. Fadel, O.B. Akan, Event-to-sink spectrum-aware clustering in mobile cognitive radio sensor networks. IEEE Trans. Mobile Comput. **15**(9), 2221–2233 (2016)
11. R. Urgaonkar, M. Neely, Opportunistic scheduling with reliability guarantees in cognitive radio networks. IEEE Trans. Mobile Comput. **8**(6), 766–777 (2009)
12. Y. Qin, J. Zheng, X. Wang, H. Luo, H. Yu, X. Tian, X. Gan, Opportunistic scheduling and channel allocation in MC-MR cognitive radio networks. IEEE Trans. Veh. Technol. **63**(7), 3351–3368 (2014)
13. S. Park, D. Hong, Optimal spectrum access for energy harvesting cognitive radio networks. IEEE Trans. Wirel. Commun. **12**(12), 6166–6179 (2013)
14. J. Ren, Y. Zhang, R. Deng, N. Zhang, D. Zhang, X. Shen, Joint channel access and sampling rate control in energy harvesting cognitive radio sensor networks. IEEE Trans. Emerg. Top. Comput. (to be published). doi:10.1109/TETC.2016.2555806
15. H. Sun, A. Nallanathan, C.-X. Wang, Y. Chen, Wideband spectrum sensing for cognitive radio networks: a survey. IEEE Wirel. Commun. **20**(2), 74–81 (2013)
16. R. Deng, Y. Zhang, S. He, J. Chen, X. Shen, Maximizing network utility of rechargeable sensor networks with spatiotemporally coupled constraints. IEEE J. Sel. Areas Commun. **34**(5), 1307–1319 (2016)
17. R. Deng, J. Chen, C. Yuen, P. Cheng, Y. Sun, Energy-efficient cooperative spectrum sensing by optimal scheduling in sensor-aided cognitive radio networks. IEEE Trans. Veh. Technol. **61**(2), 716–725 (2012)
18. M.J. Neely, *Stochastic Network Optimization with Application to Communication and Queueing Systems* (Morgan & Claypool Publishers, 2010)
19. L. Huang, M.J. Neely, Utility optimal scheduling in processing networks. Perform. Eval. **68**(11), 1002–1021 (2011)
20. G. Shah, F. Alagoz, E. Fadel, O. Akan, A spectrum-aware clustering for efficient multimedia routing in cognitive radio sensor networks. IEEE Trans. Veh. Technol. **63**(7), 3369–3380 (2014)

21. S. Boyd, L. Vandenberghe, *Convex Optimization* (Cambridge University Press, 2004)
22. L. Ramshaw, R.E. Tarjan, On minimum-cost assignments in unbalanced bipartite graphs. HP Labs technical report HPL-2012-40R1 (2012)
23. Y. Cui, V.K.N. Lau, R. Wang, H. Huang, S. Zhang, A survey on delay-aware resource control for wireless systems;large deviation theory, stochastic Lyapunov drift, and distributed stochastic learning. IEEE Trans. Inf. Theory **58**(3), 1677–1701 (2012)

Chapter 5
Conclusion and Future Research Directions

In this chapter, we conclude this book and discuss the future research directions.

5.1 Concluding Remarks

In this monograph, we have investigated the efficient utilization of harvested energy and idle licensed spectrum in ESHSNs. On the basis of the analysis and discussion given in the monograph, we present the following concluding remarks.

- We have studied the basic concept of WSN, the enabling techniques for ESH capabilities, and the need to integrate ESH capabilities into WSNs. In addition, we have introduced the architecture and applications of ESHSNs, and discussed several challenges of efficient resource utilization in ESHSNs. Furthermore, a literature survey of resource allocation in ESHSNs is provided.
- We have proposed a resource allocation solution for HSHSNs which consist of EH-powered spectrum sensors and battery-powered data sensors. The imbalance of energy replenishment and consumption at either the spectrum or data sensors results in node failure and deteriorates the network performance. To address this issue, we have designed a unified solution to schedule the spectrum sensors to detect licensed channels, and allocate the available channels to data sensors for sensed data collection. The unified solution is achieved by two algorithms that operate in tandem to guarantee the sustainability of spectrum sensors and minimize energy consumption of data sensors. Considering EH dynamics, an integer programming is formulated and addressed by the cross-entropy algorithm. To minimize the energy consumption of data sensors, we formulate a biconvex problem to jointly allocate data sensors' transmission time, power, and channels. Extensive simulation results demonstrate that, with the proposed solution, the spectrum sen-

sors can sustainably discover available channels, and the data sensors can conserve much energy in data transmission.

- We have investigated the joint allocation of energy and spectrum for ESHSNs, taking into account the stochastic nature of EH process, PU activities, and channel conditions. In specific, we have designed a network utility optimization framework to decompose the stochastic problem into three deterministic subproblems: battery management, sampling (i.e., sensing) rate control, and resource (i.e., channel and data rate) allocation on the basis of Lyapunov optimization. Under the developed framework, we have proposed an online and low-complexity algorithm which does not require any priori knowledge of the stochastic processes. To evaluate the performance of the proposed algorithm, we have analyzed the upper bounds on data queues and collision queues and revealed the required battery capacity to support the operation of the proposed algorithm. Furthermore, we have quantified the gap between the achieved network utility and the optimal network utility. Extensive simulation results demonstrate that, with the proposed algorithm, the ESHSN can achieve high network utility while guaranteeing the PU protection and network stability.

5.2   Future Research Directions

This monograph provides the preliminary results on the efficient resource utilization of ESHSNs, including the study of EH-powered spectrum sensing and energy efficient spectrum access in Chap. 3, and the network utility optimization considering the stochastic nature of EH process and PU activities in Chap. 4. In the future, we plan to investigate the resource allocation of ESHSNs based on the real data-driven EH process and PU activities modeling, to improve the practical value of the designed algorithms. Furthermore, we intend to design joint spectrum sensing and access schemes using harvested energy, and study resource allocation in multi-hop ESHSNs.

5.2.1   Real Data-Driven EH Process and PU Activities Modeling

The efficient resource utilization of ESHSNs heavily relies on the accurate models of the EH process and PU activities. The model should be as realistic as possible to facilitate the energy allocation and spectrum access decisions. Otherwise, the modeling mismatch may significantly degrade the network performance. Currently, this issue remains open mainly due to the lack of comprehensive historical data records in EH process and licensed spectrum usage. In the coming era of Internet of Things, more data records of the two processes can be complemented by the widely deployed devices with sensing capability, e.g., smartphones. Based on the records, the operator can use the data mining and machine learning techniques to extract the statistical information of EH process and PU activities.

5.2.2 Joint Spectrum Detection and Access

In practical ESHSNs, the TPS assumed in Chap. 4 may not be available due to the limitation on the deployment and maintenance cost. Without the assistance of the TPS, sensors have to jointly realize spectrum sensing and access using harvested energy. In this case, both the exploration and exploitation of licensed spectrum consume energy, which make the coupling of energy and spectrum allocation more complex. The trade-off exists between the accurate information on PU activity and immediate spectrum access.

5.2.3 Resource Allocation in Multi-hop ESHSNs

To fully cover an AoI for pervasive monitoring, some sensor networks may consist of thousands of nodes which transmit data to the sink through multi-hop relaying. Considering the signaling overhead of centralized algorithms, distributed algorithms are desired to improve the scalability of EHCRSNs. To utilize the benefits brought by EH and CR capabilities, sensors need to frequently adjust power control, change next-hop relay, and vacate operating channel according to the highly dynamic conditions of the ambient energy source and PU activities.

Printed in the United States
by Bookmasters

Printed in the United States
By Bookmasters